Rib Fracture Management

Marc de Moya • John Mayberry
Editors

Rib Fracture Management

A Practical Manual

Springer

Editors
Marc de Moya
Medical College of Wisconsin
Froedtert Trauma Center
Milwaukee, WI
USA

John Mayberry
Saint Alphonsus Regional Medical Center
Boise, ID
USA

ISBN 978-3-319-91643-9 ISBN 978-3-319-91644-6 (eBook)
https://doi.org/10.1007/978-3-319-91644-6

Library of Congress Control Number: 2018950421

Printed on acid-free paper

This Springer imprint is published by the registered company Springer Nature Switzerland AG
The registered company address is: Gewerbestrasse 11, 6330 Cham, Switzerland

Foreword

This is a timely and necessary contribution to the literature, written and edited by true experts in the field. The management of chest wall injuries has often played a secondary role in the minds of trauma surgeons. Attracted by more "flamboyant" injuries, such as internal organ lacerations and exsanguinating conditions, they have often concluded their evaluation of rib and sternal fractures with a single, routine statement about managing pain. But these injuries are so much more than that. They cause significant suffering, are indicators of other injuries, cause short- and long-term complications, and can easily be a dominant factor in an ultimate bad outcome. Relatively recently, rib fractures have been placed more prominently on the surgical radar screen due to the reemergence of interest in stabilizing them surgically. Unfortunately, the research on the topic is still suboptimal, the indications uncertain, and the techniques under evolution.

To that effect, the current book serves to clarify the confusion by offering expert opinions, supported by the existing evidence and, equally importantly, by many years of personal experience. I congratulate the editors because they did not make operative strategy the sole centerpiece of the book. As important as surgery is, it is even more important for the reader to understand the pathophysiology and consequences of chest wall injuries. It is important to learn standardized algorithms of care, but also be able to individualize therapies according to age, injury severity, and expected outcomes. By examining the issue from A to Z, from history to future directions, and by including every possible aspect of diagnosis and treatment of chest wall injuries, the authors provide the readership with an invaluable piece of knowledge, one that I suspect will be a standard of reference for many years to come.

George C. Velmahos, MD, PhD
John Burke Professor of Surgery
Harvard Medical School
Boston, MA, USA

Preface

Rib Fracture Management is a comprehensive text focused on the physiology and management of trauma-related injuries to the chest wall. The management of lung contusion, rib fractures, and chest wall dysfunction has evolved more recently as a result of the introduction of advanced mechanical ventilation strategies, extracorporeal membrane oxygenation technology, and rib fixation techniques. However, much remains unknown and continues to evolve. This text is the first to bring together chest wall injury experts from across the globe to review and consolidate evidence-based medicine for practitioners providing care for trauma patients. The content of the text was constructed in a way to provide a strong foundation in chest wall injury physiology and nonsurgical management options and covers the spectrum of injury that may benefit from surgical reconstruction using references to the most current evidence. Our intended audience is surgeons and intensivists who manage traumatic chest injuries.

I would like to acknowledge my coeditor, Dr. John Mayberry, and the contributors to the first comprehensive textbook focused on chest wall injury. We would especially like to acknowledge the staff at Springer for providing an accurate and high-quality text and in particular Margaret Burn from Springer for her devotion to ensuring the completion of this work.

I would also like to thank my wife, Adriana, and my two children, Andre and Sophia, for their support and love. Dr. Mayberry would like to thank Amy, Billie, John Jr., and Eliza for their patience during his long work hours and research endeavors.

Milwaukee, WI — Marc de Moya
Ketchum, ID — John Mayberry

Contents

Contributors

Suresh Agarwal, MD, FACS, FCCM Division of Trauma and Critical Care Surgery, Department of Surgery, Duke University Medical Center, Durham, NC, USA

Michael Bemelman, MD Elisabeth Two Cities Hospital, Tilburg, The Netherlands

Umar Bhatti, MD Acute Care Surgery, Trauma, Burn, Surgical Critical Care, Department of Surgery, University of Michigan Health System, Ann Arbor, MI, USA

Walter Biffl, MD Trauma and Acute Care Surgery Scripps Memorial Hospital, La Jolla, CA, USA

Scripps Memorial Hospital, La Jolla, CA, USA

Jose J. Diaz Jr, MD, CNS, FACS, FCCM Division Acute Care Surgery, Department of Surgery, R Adams Cowley Shock Trauma, University of Maryland School of Medicine, Baltimore, MD, USA

Andrew Doben, MD Division of Trauma, Acute Care Surgery and Surgical Critical Care, Department of Surgery, University of Massachusetts Medical School–Baystate, Springfield, MA, USA

John G. Edwards, MB, ChB, PhD, FRCS (C/Th) Department of Cardiothoracic Surgery, Northern General Hospital, Sheffield Teaching Hospitals NHS Foundation Trust, Sheffield, South Yorkshire, United Kingdom

Mario Gasparri, MD Department of Thoracic Surgery, Medical College of Wisconsin, Milwaukee, WI, USA

Marisa Gasparri University of Wisconsin – Madison, Madison, WI, USA

Krista Haines, DO Division of Trauma and Critical Care Surgery, Department of Surgery, Duke University Medical Center, Durham, NC, USA

William J. Hunt, BMedSci Department of Cardiothoracic Surgery, Northern General Hospital, Sheffield Teaching Hospitals NHS Foundation Trust, Sheffield, South Yorkshire, UK

William Long, MD Legacy Emanuel Medical Center, Portland, OR, USA

Silvana Marasco, MD Department of Cardiothoracic Surgery, The Alfred Hospital, Melbourne, Melbourne, VIC, Australia

John Mayberry, MD St. Lukes Wood River Medical Center, Ketchum, ID, USA

Tashinga Musonza, MD Michael E. DeBakey Department of Surgery, Baylor College of Medicine, Houston, TX, USA

Lena M. Napolitano, MD, FACS, FCCP, MCCM Acute Care Surgery, Trauma, Burn, Surgical Critical Care, Department of Surgery, University of Michigan Health System, Ann Arbor, MI, USA

Deepika Nehra, MD Division of Trauma, Burns and Surgical Critical Care, Department of Surgery, Brigham and Women's Hospital, Harvard Medical School, Boston, MA, USA

Raminder Nirula, MD, MPH Acute Care Surgery Section, Department of Surgery, University of Utah, Salt Lake City, UT, USA

Barbara U. Okafor, BS Division of Trauma, Burns and Surgical Critical Care, Department of Surgery, Brigham and Women's Hospital, Boston, MA, USA

Fredric M. Pieracci, MD, MPH Department of Surgery, Denver Health Medical Center, University of Colorado School of Medicine, Denver Health Medical Center, Denver, CO, USA

Richard M. Pino, MD, PhD, FCCM Department of Anesthesia, Critical Care and Pain Medicine, Massachusetts General Hospital, Harvard Medical School, Boston, MA, USA

Noelle N. Saillant, MD Division of Trauma, Emergency Surgery and Surgical Critical Care, Department of Surgery, Massachusetts General Hospital, Harvard Medical School, Boston, MA, USA

Jeffrey J. Skubic, DO Division of Trauma, Burns and Surgical Critical Care, Department of Surgery, Brigham and Women's Hospital, Harvard Medical School, Boston, MA, USA

Abraham Sonny, MD Department of Anesthesia, Critical Care and Pain Medicine, Massachusetts General Hospital, Harvard Medical School, Boston, MA, USA

S. Rob Todd, MD, FACS, FCCM Michael E. DeBakey Department of Surgery, Baylor College of Medicine, Houston, TX, USA

Thomas W. White, MD, FACS, CNSC Intermountain Trauma and General Surgery, Murray, UT, USA

Tiffany Zens, MD Department of Surgery, University of Wisconsin, UW Hospital and Clinics, Madison, WI, USA

Frank Zhao, MD Acute Care Surgery, The Queen's Medical Center, Honolulu, Honolulu, HI, USA

Department of Surgery, John A. Burns School of Medicine, University of Hawaii, Honolulu, HI, USA

History of Rib Fracture Management

1

Michael Bemelman, William Long, and John Mayberry

The Ancients

The occurrence of injury in everyday life and in battle makes it likely that ancient surgeons were familiar with chest wall injuries. Although communities were primarily rural, urban centers of several thousand people were present in Sumer (present-day Iraq) as early as 4000 BCE and in Egypt by 3500 BCE [1]. Multilevel dwellings, temples, canals, bridges, and extensive walls were built. Both the Sumerians and the Egyptians had a professional military with battalions of foot soldiers and fighting units including chariots and archers. Rib fractures, flail chest, and open chest wounds occurring during farming, construction projects, and conflicts were surely similar to what surgeons treat today [2] (Table 1.1, Fig. 1.1).

Sumerian tablets dating to 3000 BCE provide the first written descriptions of medical care [1, 3]. *Asu*, Sumerian general practitioners, used hot water, oils, wine, and honey to cleanse and dress wounds. They were aware of the risk of infection, *ummu* (hot thing). The Sumerians had medical corps accompanying their armies in the field.

M. Bemelman
Elisabeth Two Cities Hospital, Tilburg, The Netherlands
e-mail: m.bemelman@etz.nl

W. Long
Legacy Emanuel Medical Center, Portland, OR, USA
e-mail: wlong@lhs.org

J. Mayberry (✉)
St. Lukes Wood River Medical Center, Ketchum, ID, USA
e-mail: john.mayberry@idahosurgeons.net

M. de Moya, J. Mayberry (eds.), *Rib Fracture Management*,
https://doi.org/10.1007/978-3-319-91644-6_1

Table 1.1 Description of rib fractures occurring following a traumatic event 4000 years ago in Egypt

Rib no.	Right	Left
1	Butterfly fracture with one part free (force from superior)	No fractures
2	One complete fracture near sternal end	No fractures
3	Two complete fractures—one midshaft and one at the sternal end with plastic deformation	Two fractures—one complete midshaft and one incomplete near sternal end
4	Three complete fractures—one at sternal end, one near sternal end, and one near vertebral end	One fracture midshaft with piece missing
5	Three complete fractures—one sternal end, two at vertebral end creating butterfly fracture	Two complete fractures—one midshaft and one at sternal end
6	Two complete fractures—one at sternal end and one midshaft with fragment disengaged	Two fractures—one complete midshaft and one incomplete with plastic deformation on the ventral side near the sternal end
7	Two complete fractures—one at sternal end	One complete fracture midshaft
8	Two complete fractures—one at sternal end and one at the vertebral end	One complete fracture midshaft
9	Two complete fractures—one at sternal end and one at the vertebral end	One fracture with severe plastic deformation at sternal end (deformation has caused rib to bend ventrally and inferiorly)
10	One complete fracture at sternal end	No fractures
11	No fractures	No fractures
12	No fractures	No fractures

From Dupras TL, Williams LJ, De Meyer M, Peeters C, Depraetere D, Vanthuyne, Willems H. Evidence of amputation as medical treatment in ancient Egypt. Int J Osteoarchaeol. 2010;20:405–23, with permission

An *Asu* may have performed the first recorded thoracotomy, perhaps to drain an abscess or hematoma. The cuneiform etchings suggest an invasive procedure through the ribs:

> three ribs…fourth rib cut open…fluid ([3], p. 52).

The Smith Surgical Papyrus dating to 1600 BCE Egypt provides the first known mention of rib injuries in the ancient medical literature [4]. The unknown author describes rib sprains, dislocations, and open chest wounds with rib fractures but curiously does not mention simple rib fractures. Breasted (the translator of the Smith Surgical Papyrus) conjectures that the omission of simple rib fractures in the treatise may have been a scribal error ([4], p. 401).

The treatment for rib sprains and dislocations was supportive binding:

> Thou shouldst bind it with *ymrw*; thou shouldst treat afterward with honey every day until he recovers ([4], p. 461).

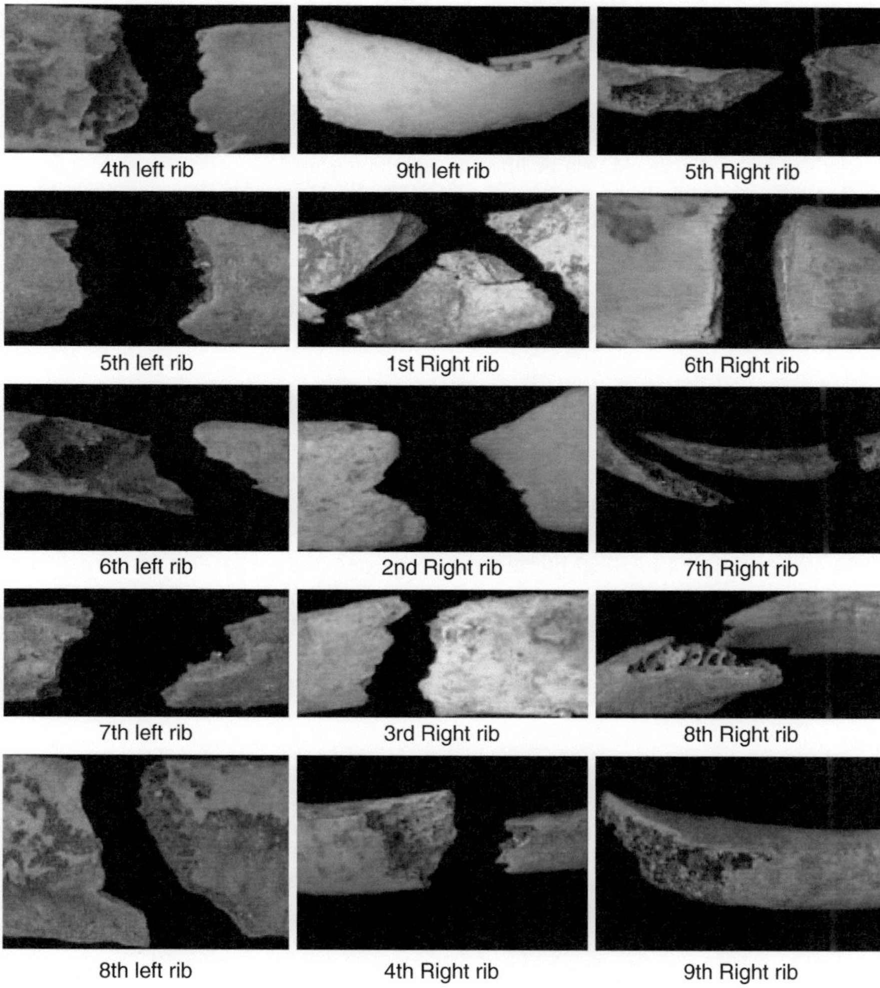

Fig. 1.1 4000-year-old rib fractures without evidence of healing found in a tomb in Dayr al-Barsha, Egypt (From Dupras TL, Williams LJ, De Meyer M, Peeters C, Depraetere D, Vanthuyne, Willems H. Evidence of amputation as medical treatment in ancient Egypt. Int J Osteoarchaeol 2010; 20:405–23, with permission)

Breasted conjectured that *ymrw* was a medication, but he found no mention of it in the Egyptian *materia medica* ([4], p. 262).

Ancient Egyptian surgeons were familiar with open chest wounds with rib fractures but did not offer any optimistic treatment strategy:

> If you examine a man having a break in the ribs of his breast, over which a wound has been inflicted; and you find that the ribs of his breast crepitate under your fingers, you should say concerning him: …An ailment not to be treated ([4], p. 462).

The Indians and Greeks

The Indian surgeon Sushruta (600–500 BCE) emphasized the splinting of rib fractures:

> In the case of a fracture of one of the *parsaka* (ribs), the patient should be lubricated with clarified butter. He should then be lifted up and the fractured rib, whether left or right, should be relaxed by being rubbed with clarified butter. Strips of bamboo ... should be placed over it and the patient should be carefully laid in a tank or cauldron full of oil with the bamboo splint duly tied up with straps of hide ([5], Vol. 2, Ch. 3).

Sushruta described *marmas* as anatomic regions that surgeons should be wary of and which when injured could lead to death ([5], Vol. 2, Ch. 6):

> A hurt to the *hridaya-marma*, which is situated in the thorax between the two breasts and above the pit of the abdomen ... proves fatal within the day.
>
> An injury to the *stana-mula-marmas*, situated immediately below each of the breasts and about two fingers in width fills the thorax with deranged (tissue), brings on cough, difficult breathing, and proves fatal.

Thus we see the first medical literature descriptions of the sequelae of sternal fracture, blunt cardiac injury, pulmonary contusion, and hemopneumothorax.

Hippocrates (400 BCE) also recommended chest wall splinting:

> ...in case of coughings, sneezing and other movements they serve as separate supports for the chest... ([6], p. 81)

Hippocrates expressed optimism for the treatment of flail chest:

> Do not make the bandaging tighter than suffices to prevent the respirations from shaking the part, or than is necessary to bring the edges of the separated fractures into touch with one another; nor is it intended to prevent coughing and sneezings, but to act as a support for the avoidance both of forcible separation and shaking ([6], p. 81).

Contemporary with Sushruta and Hippocrates were unidentified surgeons in China. Unfortunately, the vast majority of their writings have been lost, believed destroyed by an angry King after the famous surgeon Hua T'o offended him. From the sophistication of the surviving surgical descriptions, we may surmise that Chinese surgeons of 400 BCE had also developed sophisticated approaches to rib fractures, flail chest, and open chest wounds [3].

The Romans

Celsus (25 BCE–50 CE) wrote an excellent anatomic description of the twelve ribs:

> These bones are curved at their highest point, and below these they are triangular, and become gradually wider as they approach the spine. As they become wider, they become blunter. And they too at the lowest part soften into cartilage at the back and float, as it were,

> since they are unconnected with any other bone except at the top, but there they held in place by very strong muscles and sinews ([7], Book VII, Ch. 1).

Celsus taught that an incomplete rib fracture will be painful but will not cause hemoptysis and will heal within 21 days. He recommended bandaging and patience ([7], Book VII, Ch. 4). He firmly stated, however, that complete rib fractures are harbingers of infection and advocated drainage of abscesses when they occurred without delay. He recognized that a persistent draining sinus would indicate the infected rib would have to be excised ([7], Book VIII, Ch. 9, [8], p. 232).

To Soranus (78–117 CE) is attributed the distinction of being the first in the known literature to recommend an acute surgical intervention for rib fractures. He advocated excision of fracture fragments for the relief of pain:

> But if any great necessity compel us, owing to the pleura being irritated, we must divide the skin and lay bare the broken part of the rib; and then putting the instrument for protecting membranes under the rib, to prevent the pleura from being wounded, cut off properly and remove the irritating pieces of bone ([8], p. 232).

Galen (129–216 CE) also performed chest wall surgery. He described a gymnast who injured his sternum and who developed an abscess at the site 4 months later. Galen not only drained the abscess but, recognizing that the sternum was necrotic, went on to resect much of it, even exposing the heart and draining pericardial pus ([9], p. 12). Galen stated the patient would likely not have survived without this aggressive resection but recommended that surgeons not emulate his example without a thorough knowledge of chest anatomy.

Islamic Surgeons

Albucasis (Abu 'l-Qasim Khalaf ibn 'Abbas al-Zahrawi, d. 1013 CE), the Cordovan surgeon of the tenth century whose textbook *On Surgery and Instruments* was the standard in Europe for several generations, recommended setting posterior rib fractures:

> by leveling out the fracture with the fingers in any manner you can so that the form is as it should be; then apply the plaster and bind the broken bone with a splint, if necessary ([10], p. 730).

He recommended ignoring anterior fractures since they are "only contused, because they are cartilaginous." ([10], p. 730).

Albucasis further noted that if there is a "depressed fracture of the ribs, then the patient will have a vehement pain and a piercing sensation like that of pleurisy, since the bone is piercing the pleura." ([10], p. 730). For treatment he referred to the "ancients" who either filled the depression with "wool soaked in warm oil" and bandaged the chest or, following the advice of Soranus, excised the depressed portion. He warned, however, of the risk of a postoperative abscess.

For displaced sternal fractures, Albucasis recommended reduction by placing a pillow between the shoulders while the patient is supine and by manipulating the shoulders and the chest wall:

> Then apply over it a plaster and a pad and place over that a splint of thin willow board or brier or similar light wood, first wrapping it in cloth; then gently bind this upon the fractured bone several times, tying it firmly; then inspect it constantly, and whenever it loosens tighten it ([10], p. 728).

The Persian Avicenna (Ibn Sina a.k.a. the Prince of Physicians) completed his five-volume *Canon of Medicine* prior to his death in 1037 CE [11]. All five volumes have recently been translated into English. Avicenna's descriptions of chest wall anatomy and function are so advanced that his teachings could stand alone in a modern textbook ([11], p. 71–3).

Avicenna likewise recommended operative management for extreme cases:

> If the fractured rib strongly resists going back and the fractured part hits the veil and harms it, you should cut the skin until you reach the fracture. Then, place the tool that protects the peritoneum under the rib so that the peritoneum does not come out. Softly and gently cut that part of the rib bone that hits the veil and harms the veil and bring it out. Then if warm swelling does not appear, bring the cut area together and place treating ointments on it and tie it, and if warm swelling emerges, dip the pads in suitable oil and place them on it. Give medication and food that relieve the warm swelling to the patient. The patient should sleep on the side that has less pain [12].

Medieval Surgeons

Theodoric Borgognoni, an Italian surgeon of the thirteenth century, included an entire chapter on rib fractures in his *Chirurgia* [13]. To relocate depressed rib fractures, he wrote:

> ...have the patient brought to a bathtub, and after dipping his hands in turpentine, he would rub on honey, pitch or bird lime, and would place his hands, pressing on the spot where the break was, and suddenly lift up, and do this repeatedly until the rib returned to its proper place ([13], Book 2, p. 191).

He also described a limited thoracotomy for blunt injury where "the bone has punctured the diaphragm":

> ... then it is necessary to cut at the point of injury and disclose the broken rib. Protect the pleura by placing an instrument under the rib so that it cannot harm it. Then it is easy to excise that part of the bone which is puncturing the pleura; afterwards follow the regular treatment for wounds ([13], Book 2, p. 191–2).

Guy de Chauliac of fourteenth-century France provided what may be the first known description of serial irrigation of the pleural space to prevent empyema ([14], p. 339). He recommended enlarging a penetrating thoracic wound to encourage pleural drainage and then described daily instillation of warm wine or dilute honey with rotation of the patient and subsequent observation of the effluent. When the fluid became clear, the irrigations were discontinued.

Hieronymus Brunschwig of Germany published *Das Buch der Cirurgia: Hantwirckung der Wundarztny* in 1497 [15, 16]. From a 1525 English translation of the section entitled "The brekking of the ribbes":

> E shall knowe that ther is ribbes longe & shorte. And the shorte ribbes breke not nyghe by the backe. Ye other ribbes breke in many places and they be sotime croked & bowed inward and is not broke. And somtyme outward also. And somtyme ye fracture is deedly by loge endurynge of payne and sotyme it is not deedly and shortely heled. And this ye may knowe hereafter written, the first ye must with your hade fele the broke place. And yf there be ony crackynge then is it broke. And yf there benone euyll accedence to se it is good to helpe. And as they bowe inward so be they euyll to helpe. And yf the ribbes be ferre sonken in that mebres be sore wounded inward it is deedly or el les longe sykenes. And that may be knowe by the short brethe and by blood spyttynge coghynge with the styche and payne in the side [15].

Brunschwig reiterated the teachings of Albucasis and Theodoric advising attempts at manual reduction with the fingers if possible, with adhesive plasters if necessary, followed by the placement of plaster wraps. For comminuted fractures he repeated the advice of Soranus: asserting an incision followed by removal of rib fracture fragments may be necessary.

Ambroise Paré of the sixteenth century wrote several descriptions of severe chest wall injuries [17]. Following a battle between the French and the Spanish, he treated a man shot by an *arquebus* where the ball passed through the chest and created an open chest wound with comminuted rib fractures:

> I saw he cast blood out of his mouth and his wounds. Moreover he had a great difficulty of breathing, and cast out wind by the said wounds with a whistling, in so much that it would blow out a candle, and he said he had a most sharp prickling pain at the entrance of the bullet. I do believe and think it might be some little pieces of bones which pricked the lungs. When they made their systole and diastole, I put my finger into him; where I found the entrance of the bullet to have broken the fourth rib in the middle and scales of bones which the said bullet had thrust in, and the outgoing of it had likewise broken the fifth rib with pieces of bone which had been driven from within outward. I drew out some but not all, because they were deep and very adherent ([17], p. 55–6).

Paré dressed the open chest wounds with linen gauze soaked in egg yolks, turpentine, and oil of roses. His dressing, he stated, allowed for the "flux of blood" but did "hinder that the outward air did not enter into the chest." He "bound him up, but not hard, to the end he might have easy respiration."

> And as for the pain which he said he felt at the entrance of the bullet … that was because the lungs by their motion beat against the splinters of the broken rib. Now the lungs are covered with a coat coming from the membrane called pleura, interweaved with nerves of the sixth conjugation from the brain, which was cause of the extreme pain he felt; likewise he had a great difficulty of breathing, which proceeded from the blood which was spilled in the thorax, and upon the diaphragm, the principle instrument of respiration, and from the laceration of the muscles which are between each rib which also help also to make the expiration and inspiration ([17], p. 57).

Eighteenth- and Nineteenth-Century Surgeons

In 1702, John Moyle, in *One of Her Majesty's Ancient Sea-Chirurgions*, wrote (italics his):

> If the Rib is broke or difplaced you must lay the Patient on the well fide over fome round Substance, as the Bilge of a Cask, or a Gun (his Clothes being off) and let your Affistant bend his Body downwards and fo will extention be made, that fo with your hands you may place the ends of the Rib together. But if it is diflocated and will not be reduced this way, then do this.
>
> Lay on the Part the Stitch Plaster, made of *Maftich, Gypfum, Terib.* and *aviarium Gluten* mixed, fowing Tape to the outfide of it to haul by when you have occafion; and when it hath lain on long enough fo as to stick fast, then lay the Man in the pasture as before,and hale up forcibly by the strings, and the Rib will come into his Place.
>
> Then Embrocate with *Ol. Rof.* and apply the *defenfive minor*, as in the Catalogue, and make decent Rouling; and at last *Emplast. Catagmatic.*
>
> But be fure here to let Blood, for this is abfolutely convenient, and give the Man *Spruce Beer* or the *Traumatic's* which you will find in the Catalogue; and let him have a breathing fweat, to hinder the Coagulation of the Blood, and heal inwardly, and let him have the frefh and wholfome Diet [18].

In 1743, Heister, the famed German surgeon, reinforced previous recommendations for manual reduction manually and surgical removal of rib fragments that irritated the pleura ([19], Vol. 1, p. 124–5). Like many trauma and thoracic surgeons of today, Heister believed retained hemothorax was at high risk of empyema and recommended preemptive drainage ([19], Vol. 1, p. 124–5).

Heister's descriptions of bandaging of the chest for clavicle, scapular, sternal, and rib fractures were meticulous. He recommended continuing the bandaging onto the lower chest "til the whole disordered part of the thorax is thus invested" ([19], Vol. 2, p. 313). In cases where the rib fracture(s) required reduction, he advised adding a soft splint and/or plaster to the bandaging.

Guthrie and Larrey, English and French military surgeons of the Napoleonic Wars, respectively, wrote extensively of the treatment of chest injuries. Both recommended incision into severe blunt and simple penetrating chest wall injuries for removal of bone splinters ([9], p. 194).

In 1830, John Hennan of the University of Edinburgh wrote:

> In every injury of the chest a firm elastic bandage is an indispensible assistant in the cure; the motions of the ribs are not only restrained, but the parts are powerfully supported by its application; if fracture has taken place in many of the bones, we have no other means so perfect of retaining them in their place… ([9], p. 196).

In 1830, William Lawrence of St Bartholomew's Hospital in London wrote:

> Fractures of the ribs are much more common than those of the sternum. When these take place at the anterior part, or sides of the chest, the accident is generally easily recognizable by putting the hand where the violence has been received, or where the patient says there is considerable pain. The movements of the chest produce a sensible grating, or crepitus, and the patient experiences great pain from the motions of the broken ends of the bone in the chest.

> Treatment – If the chest could be kept perfectly at rest – if the patient did not employ the intercostal muscles at all, there would be no movement of the fractured ends of the bones, and no material pain … We endeavor to accomplish this as well as we can, by covering the part, either by a broad bandage of calico or flannel, or by including it in a broad kind of girth, fasten, fastened with buckles and straps, called a fractured-rib bandage.
>
> In some instances, however, the pressure of this bandage, and the swelling that takes place, seem to act unfavourably on the broken ends of the bone, and to aggravate the sufferings of the patient; and this is particularly the case where several of the ribs are broken; so that we often find it necessary to leave the chest without external pressure… ([9], p. 213).

In 1861, Tripler and Blackman, in *The Handbook for the Military Surgeon*, advised:

> Broken ribs, from gunshot wounds, sometimes gives rise to persistent inflammations and empyema, and require the wound to be enlarged to let out the matter, pieces of bone, clothing, etc.
>
> … pick away carefully the broken pieces of bone, to pare off the ragged edges of the rib, to apply proper bandages to the fracture, and to keep the patient lying on that side. The principles of the treatment of this fracture do not differ from those that govern the rib fracture from any other cause ([20], p. 78–9).

The authors of *The Surgical History of the War of the Rebellion* indicate that excision of comminuted rib fractures following penetrating wounds of the chest during the American Civil War was not unusual ([21], p. 566–70).

Awakenings

In 1928, Holderman published a history of the management of sternal fractures [22]. He noted that available textbooks stated that surgical reduction followed by wiring or plating was occasionally justified but was prone to failure. He advocated manual reduction, the placement of plaster swaths around the upper chest, and bed-rest to keep the sternal reduction in place.

In 1942, DeBakey recommended intercostal nerve blocks and partial chest wall strapping with adhesive tape to help control pain and restore respiration in cases of "crushed chest" [23]. He described manual reduction of displaced rib fractures as "unnecessarily complicated and perhaps exhausting."

In 1942, Gray, describing what appears to be a patient with flail chest, recommended the placement of "light sandbags" or "the use of a respirator" in cases where the rib is fractured "fore and aft" and the chest wall is "not stable" [24].

In 1942, Leavitt reported a patient with multiple rib fracture nonunions persisting 1 year after a fall [25]. Leavitt harvested cortical tibial grafts and fashioned them into "shuttles" that he inserted into the "freshened" rib medullary canals. Two of the grafts did not heal, and a second surgery was required in which he "wired" the persistent nonunions successfully.

In 1943, Elkin and Cooper stated, "If bony fragments are depressed and are resulting in pleuropulmonary damage, these will require elevation" [26].

In 1947, Richardson and Papper of Bellevue Hospital, New York, citing reports of tissue necrosis following alcohol injections of the intercostal nerves, could not recommend its use for rib fracture pain relief, in spite of earlier reports of its benefits by Italian surgeons [27]. They reported their results of a series of 217 rib fracture patients treated with paravertebral and intercostal nerve blocks with procaine. Eighty percent of patients had "good" results, and only 25% required repeat blocks. The authors noted that the local anesthetic blocks obviated the need for opioids and chest wall strapping/splinting.

In 1950, Coleman and Coleman of the Medical College of Virginia, referring to the mention by Hagen in 1945 of "wiring" rib fractures [28] and of Shefts and Doud in 1946 of "wiring" following traumatic thoracotomies [29], reported 15 patients with flail chest for whom they wired transverse and oblique rib fractures in two planes and in select cases of transverse fracture carved a 3 cm "peg" of the bone from an adjacent rib and placed it through drill holes on either side of the fracture [30] (Fig. 1.2).

In 1951, Carter and Giuseffi of Cincinnati, recognizing the detriments of chest wall strapping on respirations and the ineffectiveness of intercostal nerve blocks for many patients, presented a case series of seven patients receiving a prophylactic tracheostomy for "crushed" or "flail" chest [31]. They were impressed by the beneficial effects of tracheostomy including enhanced secretion clearance, lessened respiratory effect, and improved oxygenation. They also noted that in several patients, the paradoxical motion of the chest wall disappeared or dramatically decreased immediately following the tracheostomy.

In 1952 Valle, a thoracic surgeon at Tokyo Army Hospital during the Korean War reported that 2% of his patients had "crush injuries" [32]. He recommended 45-degree angle head elevation, frequent coughing, minimal sedation, sparing opioids, and intercostal nerve blocks with 2% procaine. He further stated, "When more than two ribs in one hemithorax were fractured, they were immobilized in the usual manner with adhesive tape." When fractures were comminuted and especially when bone fragments were "driven into the pleural space and lung parenchyma," he stated the fragments were removed and the "edges trimmed" at thoracotomy.

In the discussion that follows Valle's presentation at the American Association for Thoracic Surgery, Dr. Frederick Harper of Denver espouses his opinion that "... some patients with flail chest...can be handled better by open thoracotomy." Dr. Harper does not describe what he would do during the thoracotomy, but as evidence

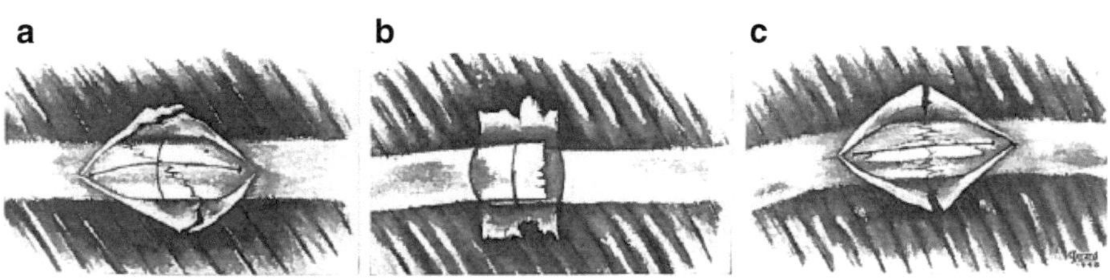

Fig. 1.2 Coleman and Coleman's description of one of the first rib fracture fixations in the modern era (From Coleman FP, Coleman CL. Fracture of the ribs—a logical treatment. Surg Gyn Obstet 1950;90:129–34, with permission)

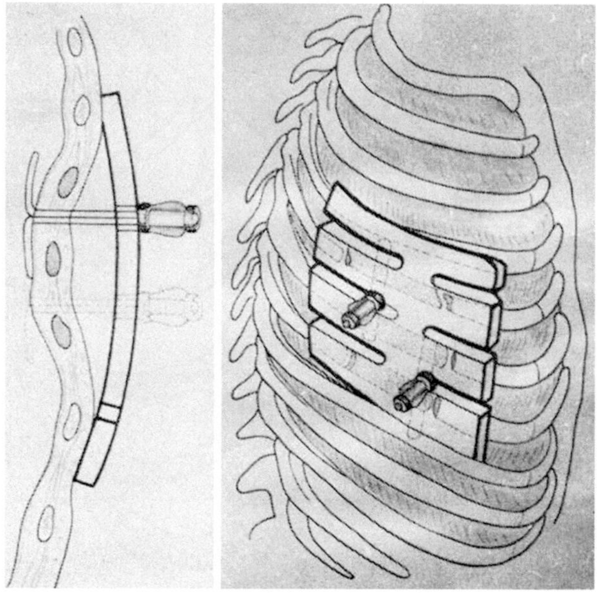

Fig. 1.3 Constantinescu's description of an ingenious and effective external rib fracture fixation plate (From Constantinescu O. A new method of treating the flail chest wall. Am J Surg 1965;109:604–10, with permission)

that "wiring" of the fractures would be considered, Dr. George Summer of Trenton, New Jersey, who followed Dr. Harper in the discussion, does mention "wiring" in a description of closing thoracoabdominal exposures although he stated, "To wire such ribs to avoid the unstable thoracic wall will undesirably prolong the operation time" [32].

In 1956, Crutcher and Nolen presented a case series of 14 patients with depressed rib fractures treated operatively with intramedullary Rush nails and wire cerclage [33]. Good results were reported with few complications.

Many varieties of external traction and fixation devices were developed for chest wall injuries in the twentieth century [34]. One of the most ingenious and least troublesome of these devices was the "panels" of Constantinescu of Bucharest, Romania, that reduced and held in place depressed chest wall segments [35] (Fig. 1.3).

Mechanical Ventilation Slows Rib Fracture ORIF Development

In 1956, the introduction of positive pressure ventilation, either via mask or endotracheal insufflation, for "critically crushed chests" by Avery, Mörch, and Benson of the University of Chicago dramatically slowed development of rib fracture operative reduction and internal fixation (ORIF) techniques [36].

Scattered reports of sternal and rib fracture ORIF occurred over the subsequent two decades but were largely ignored.

In 1957, Henry reported a dramatically positive effect from plating an unstable sternum in a case of "stove in chest" [37].

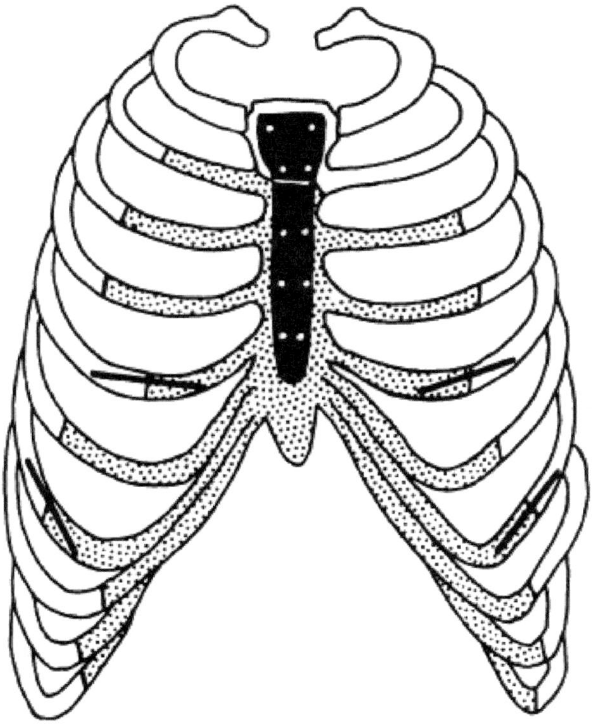

Fig. 1.4 Sillar's description of intramedullary costo-sternal fixation and sternal plate for anterior flail chest (From Sillar W. The crushed chest: Management of the flail anterior segment. J Bone Joint Surgery 1961;43B:738–45, with permission)

In 1961, Sillar of Glasgow, Scotland, reported using intramedullary wires for anterior flail segment with the addition of a malleable sternal fixation plate if needed [38] (Fig. 1.4).

In 1971, Le Roux and Stemmler of South Africa reported the use of stainless steel "Adkins" struts in the correction of a "stove in chest" that occurred in a patient who fell from a building [39].

In 1973 and 1974, reports of rib internal fixation struts introduced in France and Spain surfaced [40–43].

In 1975, from the United Kingdom, Moore presented an enviable series of 50 patients with blunt chest wall injury stabilized with intramedullary pins [44].

Also in 1975, in the United States, Richardson, Grover, and Trinkle recommended early operative wiring of isolated sternal fractures, [45] and Thomas, Blaisdell, Lewis, and Schlobohm, citing their experience that the duration of mechanical ventilation "is frequently prolonged when extreme chest trauma is present," advocated for selective operative fixation of flail chest [46].

The Germans were not inactive: "Rehbein" intramedullary nails were described in 1978, "Vecsei" plates in 1979, and "Labitzke" plates in 1981 [47–49]. In 1981, Hellberg and colleagues of the University of Göttingen described the use of compression osteosynthesis plates (normally used for facial fracture fixation) for flail chest [50].

The Recognition That Rib Fractures Are a Marker for Non-thoracic Injuries

The anticipation that the respiratory status of patients with rib fractures will worsen over the first several days is a clinical precept that dates back to Sushruta and Hippocrates. The recognition that rib fractures are a "marker" for other injuries and for mortality was not established, however, until the 1990s. Three publications are largely responsible for this step forward. Lee and colleagues analyzed the Maryland state trauma database for 3 years 1984–1986 (105,683 patients) and found that the presence of three or more rib fractures identifies a subset of patients likely to require tertiary care [51]. In 1990, Garcia and colleagues of the Children's National Medical Center showed that the risk of mortality in children dramatically increases with the number of ribs fractured [52]. In 2000, Bulger and colleagues at Harborview Medical Center in Seattle showed that this concept was also true for the elderly (65 years and older) [53].

Resurgence of Chest Wall ORIF

The recent surge of rib fracture ORIF began in the early 1980s in Portland, Oregon, USA, with Dr. William Long and colleagues of Legacy Emanuel Hospital (LEH). Long had completed a trauma and surgical critical care fellowship at the University of Maryland Shock Trauma Center in Baltimore followed by cardiothoracic surgery training at the University of California, San Diego. During his entire training at Baltimore and San Diego, Long had never seen nor heard of rib fixation as a way to treat flail chest.

In 1983, Long encountered a 30-year-old lady who had been pinned against a brick wall by a truck, crushing her left anterolateral chest and tearing her left upper lobe (LUL) bronchus from the trachea. She was taken directly to the operating room for a left anterolateral thoracotomy with a LUL bronchus reanastomosis. Her left anterior 3–7 ribs had comminuted fractures and were very unstable. Conventional treatment at that time was to leave the patient intubated and support her chest wall with positive pressure ventilation for 3–4 weeks. Long was loath, however, to leave her on a ventilator for 3–4 weeks with a fresh bronchial anastomosis.

Long and Dr. Robert Wilson, who had recently completed a trauma orthopedic fellowship at the University of Michigan, passed Luque wires around the fractured ribs, which were then pulled into alignment and secured with vertical plates, thus suspending the flail segment like a suspension bridge (Fig. 1.5). The patient was extubated in 2 days and discharged home in 4 days. Three months later, she returned to work in the kitchen at LEH. Six months postoperatively, however, she requested that the plates be removed because she felt they were "too heavy."

Over the ensuing years, Long and colleagues plated several additional patients with severe flail chest, including those with "stove in chest" requiring a "traumatic

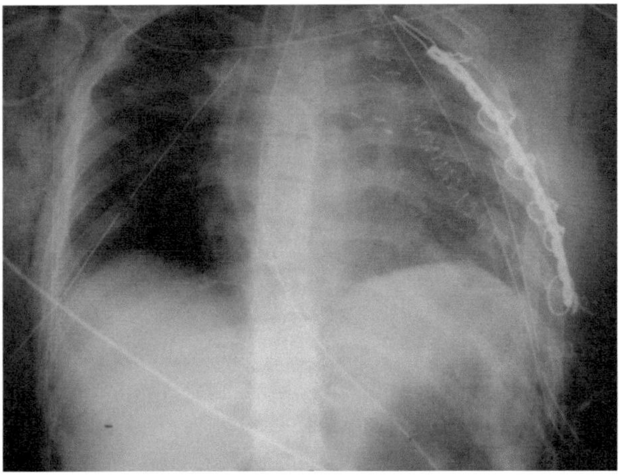

Fig. 1.5 Vertical plate reconstruction of crushed chest (Courtesy of W. Long)

thoracoplasty" with minimal to no complications. In several cases, however, patients again requested plate removal because of a sensation of feeling, e.g., "out of plumb."

In 1985, Long presented his experience with vertical rib plating in a poster display at the American College of Surgeons Clinical Congress and received both interest and disapproval. Prominent trauma surgeons reviewing his poster were concerned, however, that vertical plating would cause restriction of lung function and chest wall mobility. Long thus worked with the LEH nuclear medicine department on a comparative study of normal lung and uninjured chest wall with contralateral flail without significant pulmonary contusion with inhaled radioactive xenon, pre- and postoperatively. The results were dramatically encouraging. While in the three injured patients studied, 80% of the inhaled xenon went to the uninjured side's lung preoperatively; after vertical fixation, there was equal distribution of counts on both sides (Fig. 1.6). Coincidentally, xenon came under restriction for human investigation, and thus a full study was not completed, and this information was never published.

In 1985, Long and colleagues treated their first patient with an unusually large acute chest wall pulmonary hernia, a middle-aged woman with a crushed left chest following a motor vehicle crash. She exhibited marked distraction of ribs # 4–8, allowing her left upper lobe to herniate through pectoral and serratus anterior muscles into her subcutaneous space, creating a large pseudobreast. Following debridement of devitalized tissue and placement of several transverse plates, the remaining 15 × 25 cm gap in the chest wall was covered with Vicryl mesh (Fig. 1.7). She returned to normal activity but, at 1 year, also requested removal of the plates.

The repetitive patient requests for removal of the pelvic reconstruction plates led Dr. Jon Hill and Long to switch to the preferred use of titanium mandibular fixation plates with bicortical screws after which the requests for plate removal by the patients ceased.

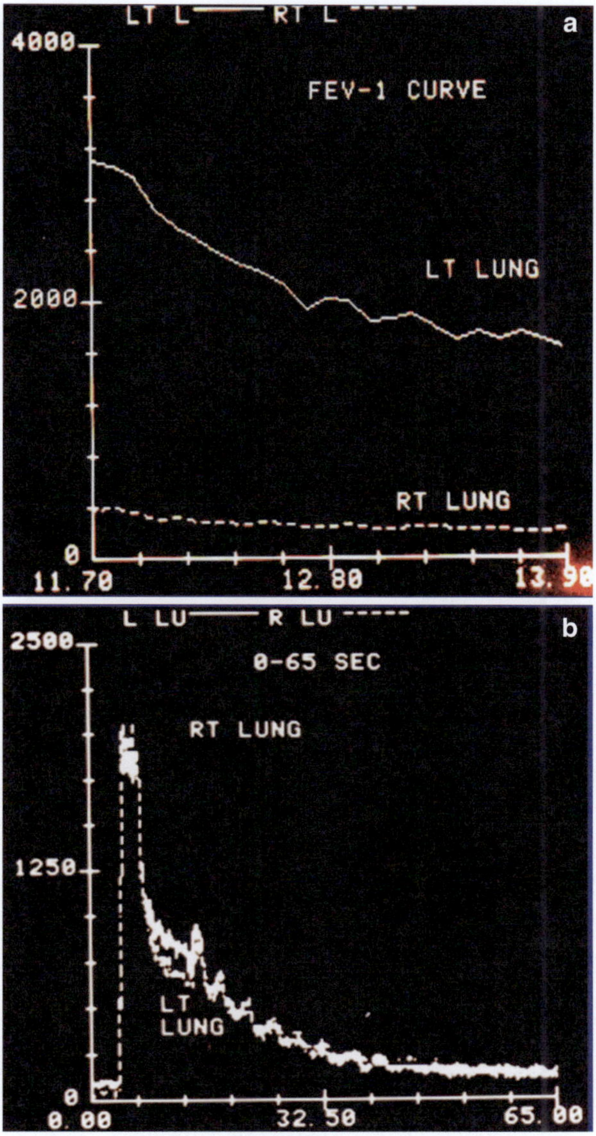

Fig. 1.6 (**a**) Preoperative xenon inhalation showing the disparity between the uninjured and injured lungs. (**b**) Postoperative xenon inhalation on the same patient (Courtesy of W. Long)

During this time, two LEH trauma orthopedic surgeons (Dr. James Krieg and Dr. Steve Madey) with the assistance of Michael Bottlang, PhD of the Legacy Biomechanics Laboratory, began designing low-profile, rib-specific, flexible titanium plates [54] (Fig. 1.8). The technology was licensed to the Synthes Corporation and has subsequently been marketed as the MatrixRIB™ (depuysynthes.com).

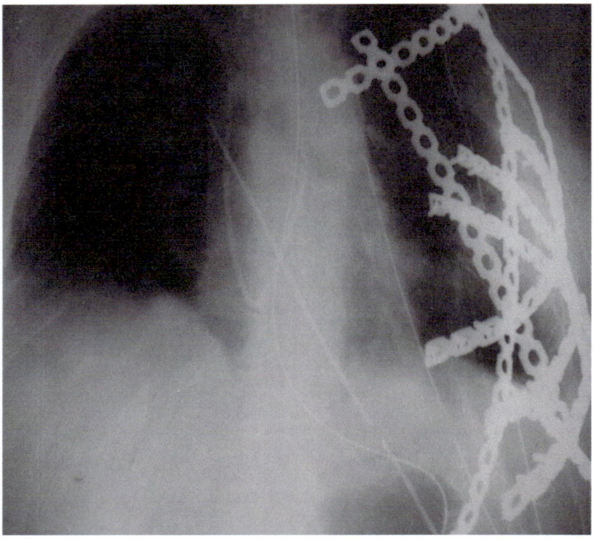

Fig. 1.7 Postoperative radiograph of patient with large acute traumatic pulmonary hernia treated with plate operative reduction and internal fixation (ORIF) and absorbable mesh (Courtesy of W. Long)

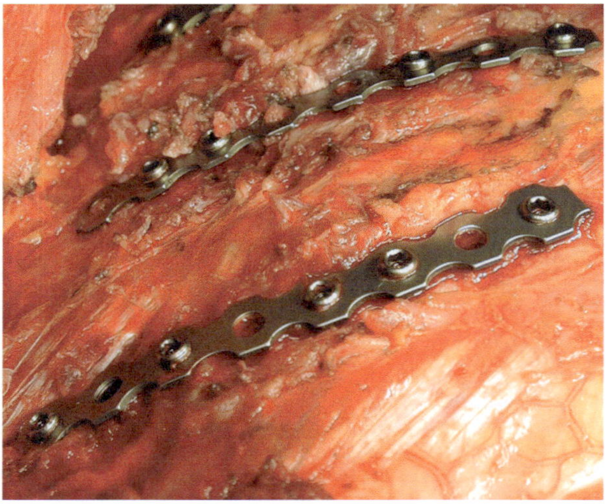

Fig. 1.8 Prototype of low-profile, rib-specific, flexible titanium plates (Courtesy of W. Long)

In 1990, Dr. George Haasler of Milwaukee, Wisconsin, USA, reported dramatic improvement of respiratory function in a 66-year-old woman with multiple, severely displaced posterolateral rib fractures following ORIF [55]. Haasler became a strong advocate for early operative stabilization along with his colleagues at the Medical College of Wisconsin, Drs. Mario Gasparri and William Tisol [56].

In 1992, Dr. Donald Trunkey performed the first flail chest ORIF at Oregon Health & Science University (OHSU) after having earlier observed rib fracture ORIF cases at San Francisco General Hospital [46]. Trunkey, along with Dr. Richard Mullins, the Trauma Section Chief, encouraged Dr. John Mayberry, Trauma Fellow, to investigate flail chest ORIF as a research endeavor. Their goals were to seek the most effective way to repair severe rib fractures, to establish selective indications, and to report long-term outcomes that could be compared to long-term nonoperative outcomes. The high rate of long-term disability of patients with flail chest treated nonoperatively reported by Landercasper and colleagues from La Crosse, Wisconsin, USA [57], greatly influenced their thinking. The trauma group at OHSU and concurrently W. Long across the Willamette River in Portland, Oregon, and H.B. Ris of Bern, Switzerland, theorized that flail chest ORIF could provide not only acute benefits but also predicted that this controversial procedure would prevent long-term pain and disability [58, 59].

In 1995, Ahmed and Mohyuddin of the United Arab Emirates repaired 26 patients with flail chest equivalents with intramedullary wires [60]. Compared with contemporary controls, these patients required less ventilator time, less tracheostomies, and less pneumonia and had lower mortality.

In 1996, Tanaka and colleagues presented a prospective randomized trial of "severe flail chest patients" treated with either early ORIF with Judet struts or internal pneumatic stabilization at the American Association for the Surgery of Trauma in Houston, Texas, with the same positive results and a dramatic return to work advantage for the stabilized patients [61].

In 1998, Voggenreiter and colleagues of Essen, Germany, presented their similar series and outcomes, although they used Judet staples or 3.5 mm anterior fixation plates [62]. From their groupings of patients comes the admonition that patients with pulmonary contusion may not benefit from flail ORIF.

In 2000, Dr. Thomas Ellis and Joel Gillard, at the OHSU Orthopedic Biomechanics Lab, devised the U-shaped/locking screw concept for rib-specific fixation (Fig. 1.9). The patent for this new prosthesis, originally called the "U-plate," was filed in 2003, and in 2005 the US Food and Drug Administration (FDA) rib fracture ORIF marketing clearance was issued to Acumed, LLC of Hillsboro, Oregon. That same year Mayberry performed the first clinical use of the "U-plate" at OHSU [63]. The biomechanical utility of the "U-plate" (now called RibLoc™, acuteinnovations.com) was demonstrated in 2008 [64].

In 2004, a case series demonstrating the utility of rib fracture ORIF in neonatal foals was presented [65]. Recognizing that newborn foals with rib fractures and

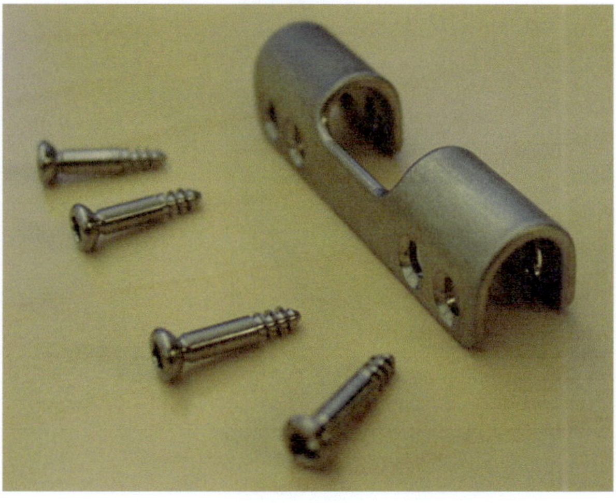

Fig. 1.9 Prototype of acute innovations, LLC rib-specific plate (Acute Innovations, LLC, Hillsboro, OR)

respiratory distress have a high mortality (greater than 50%) and after reviewing the literature showing the benefits of human rib fracture ORIF, these surgeons at the Tufts University School of Veterinary Medicine trialed equine rib fracture ORIF. Of 14 foals (age 1 h–5 days) that presented with respiratory distress and underwent rib fracture ORIF, 12 were discharged in good condition. There was one death due to migration of a Steinmann pin into the foal's heart. Following this incident, the authors used 2.7 mm plates and screws provided by the Synthes Corporation and had no further serious surgical complications.

In China, unique operative techniques and instruments have been developed in the past decade. Due to the relative isolation of China and the language barrier, these techniques are relatively hidden from the rest of the world. Chinese surgeons have been using clips made from nickel and titanium (nitinol), a shape memory alloy, for several years (Fig. 1.10). This alloy has two unique features: the capability of shaping back to its original form and size and enormous elasticity, ten times greater than conventional alloys [66]. Nitinol's elasticity may be a valuable aspect for use in rib fixation considering that ribs move approximately 14,000 cycles/day with respiration.

The Chinese nitinol clips are based on the same principle as the Labitzke and Stracos plates [49, 67]. The plates are chilled with sterile ice water, the size is chosen, the hooks are bent open to facilitate placement, and the plate is rapidly applied. As the plate warms to body temperature, it will return to its original form. These clips may have benefit in anatomical regions difficult or impossible to access, e.g., under the scapula (Fig. 1.11). The system is not suitable, however, for large comminuted fractures or fractures next to the spine or sternum. Hardware removal can also be a challenge; when the surgeon attempts to bend open the hooks, they immediately want to go back to their original form; therefore quite often the plate has to be cut.

Regretfully no nitinol clip rib fracture ORIF clinical data exist to our knowledge. In discussions with Prof. Wu Jun of Beijing, one of the authors of this review (MB), has learned that the workhorse rib fixation technique in China is conventional plate and screw fixation supplemented with nitinol clips for simple fractures and those that are difficult to access.

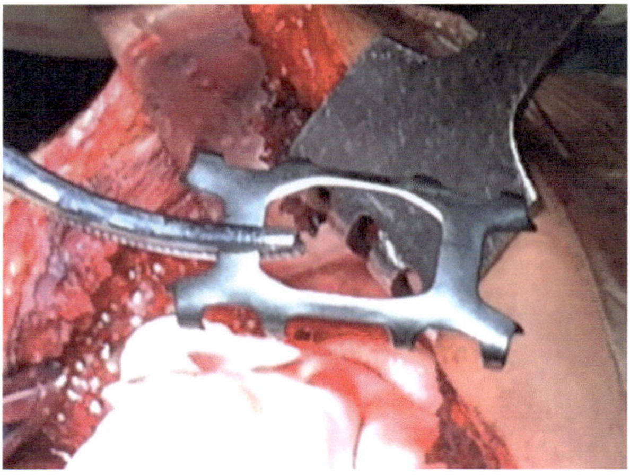

Fig. 1.10 Chinese rib-specific nitinol memory plate (Courtesy of Prof. Wu Jun, Beijing, China)

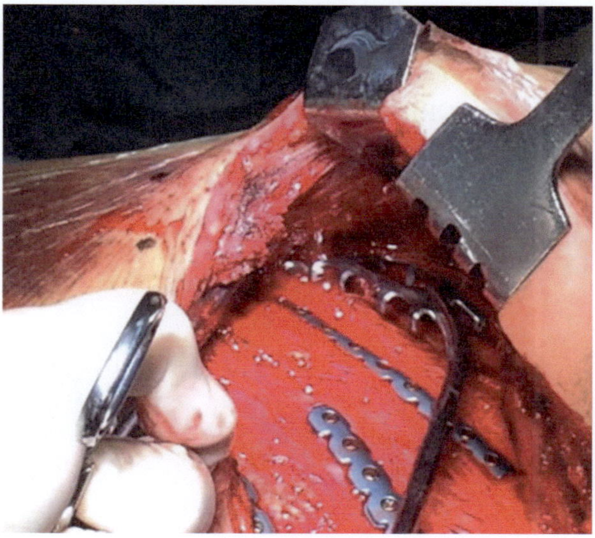

Fig. 1.11 Facility of Chinese rib-specific nitinol plate for subscapular placement (Courtesy of Prof. Wu Jun, Beijing, China)

Acceptance of ORIF

In 2010, the National Institute for Clinical health and clinical Excellence (NICE), UK, issued a guideline for rib fixation [68]. The guideline stated that with current evidence, fixation with flail chest patients can be offered routinely, provided the patient understands the procedure and agrees; and the results of the procedure should be monitored. Seminal supporting studies subsequently reported include a decision analysis led by Dr. Ram Nirula of the University of Utah, Salt Lake City, USA [69]; a randomized trial led by Dr. Silvana Marasco of The Alfred Hospital, Melbourne, Australia [70]; a study of ORIF "rescue therapy" led by Dr. Andrew Doben of Baystate Medical Center, Springfield, Massachusetts [71]; and a series of severely injured patients benefiting from "preemptive" surgical intervention led by Dr. Fred Pieracci of the University of Colorado [72].

Minimally Invasive Approach

In 1975, Paris and colleagues of Centro Hospitalario La Fe de Valencia, Spain, were the first surgeons to describe a minimally invasive approach to flail chest ORIF. Their technique included inserting a plate percutaneously between two small incisions [42].

In 1998, Tagawa and colleagues of Kitakyushu Yahata City Hospital of Fukuoka, Japan, reported three patients in whom they performed a video-assisted thoracoscopic approach combined with a mini-thoracotomy to reduce and fixate rib fractures with absorbable polylactide plates or polypropylene mesh [73].

In 2002, Sing and colleagues at the Carolinas Medical Center in Charlotte, North Carolina, made history when two millennia after Soranus described open resection they reported thoracoscopic resection of intruding rib fracture fragments for the relief of pain [74].

In 2006, Mayberry and colleagues presented a video of a thoracoscopic-assisted rib fracture ORIF with absorbable plates at the American College of Surgeons Trauma Surgery Video Session [75].

The Chinese have developed a nitinol plating system for a thoracoscopic approach (Fig. 1.12). In a reverse fashion, the plate is introduced in the chest cavity

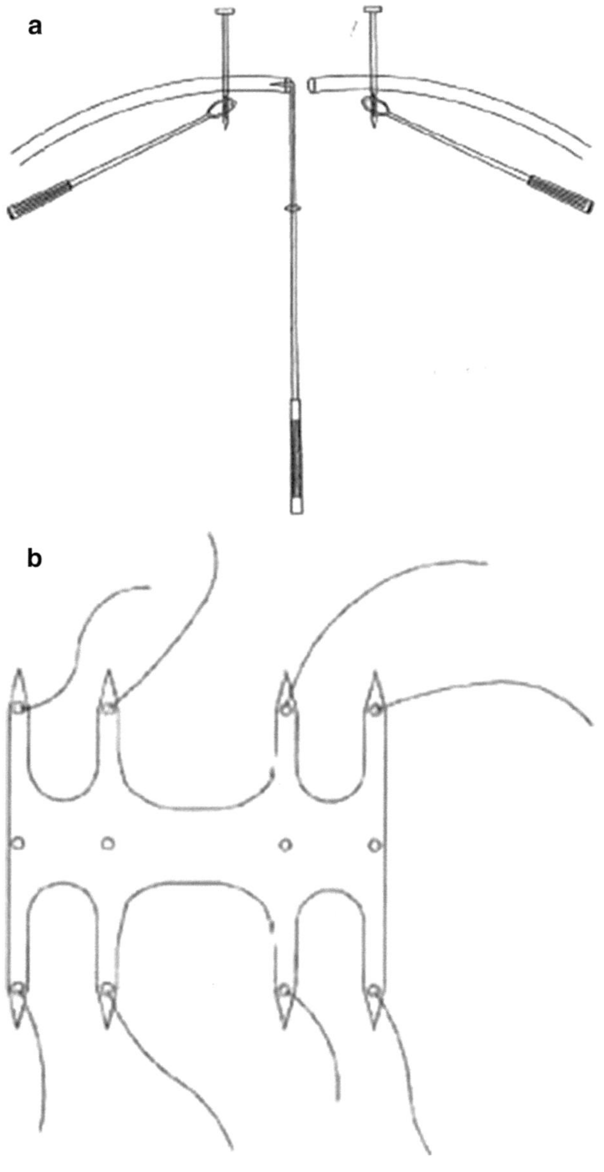

Fig. 1.12 (**a** and **b**) Wu Jun's description of thoracoscopic approach to rib fracture ORIF using nitinol memory plates (Courtesy of Prof. Wu Jun, Beijing, China)

after localizing the rib fracture and controlling it with small pins, which are drilled through the ribs. Additional small loops are provided to reduce the fracture thoracoscopically. The plate is threaded with sutures and applied underneath the rib. The wires are passed through the soft tissue and wrapped around the rib, thus stabilizing the fracture.

In 2015, Pieracci and colleagues of Denver, Colorado, USA, reported a completely thoracoscopic rib fracture ORIF [76]. In 2016, Bemelman and colleagues of Utrecht, the Netherlands, described minimally invasive plate osteosynthesis (MIPO) of rib fractures utilizing three-dimensional CT scan images, thoracic landmark recognition, preoperative rib fracture ultrasound, specialized wound retractors and rib clamps, a 90 degree drill and screwdriver, a threaded reduction tool, and percutaneous fixation [77].

References

1. Gabriel RA. Man and wound in the ancient world. Washington, D.C.: Potomac Books; 2012.
2. Dupras TL, Williams LJ, De Meyer M, Peeters C, Depraetere D, Vanthuyne B, Willems H. Evidence of amputation as medical treatment in ancient Egypt. Int J Osteoarchaeol. 2010;20:405–23.
3. Majno G. The healing hand: man and wound in the ancient world. Cambridge, MA: Harvard University Press; 1975.
4. Breasted JH. The Edwin Smith surgical papyrus, translation and commentary, vol. 1. Chicago: The University of Chicago Press; 1930.
5. Bhishagratna KKL. An English translation of the Sushruta Samhita, vol. 2. Calcutta: Chowkhamba Sanskrit series Office; 1911.
6. Jones WHS. Hippocrates, English translation, vol. III. Cambridge, MA: Harvard University Press; 1957.
7. Spencer WG. Celsus De Medicina, English translation. Cambridge, MA: Harvard University Press; 1961.
8. Hurt R. The history of cardiothoracic surgery from early times. New York: Parthenon Publishing Group; 1996.
9. Hochberg L. Thoracic surgery before the 20th Century. New York: Vantage Press; 1960.
10. Spink M, Lewis G. Albucasis on surgery and instruments. Berkeley and Los Angeles, CA: University of California Press; 1973.
11. Baktiar L. The Canon of medicine, Avicenna, 1, English Translation. Great Books of the Islamic World, Inc., KAZI Publications, Inc., Chicago, 1999.
12. Baktiar L. The Canon of medicine, vol. IV. Chicago: English Translation. Great Books of the Islamic World, Inc., KAZI Publications, Inc.; 2017.
13. Campbell E, Colton J. The surgery of theodoric, volume one. New York: Appelton-Century-Crofts, Inc; 1955.
14. Rosenbaum L. Guy de Chauliac Chirugia Magna, An English translation. Bloomington: Xlibris Corporation; 2007.
15. Treves P. 'Handy Warke of Surgeri', English Translation of Brunschwig's Das Buch der Cirurgia: Hantwirckung der Wundarztny, 1525, Early English Books Online, eebo.chadwyck.com, Accessed June 2017.
16. Di Matteo B, Tarabella V, Filardo G, et al. The traumatologist and the battlefield: the book that changed the history of traumatology. J Trauma Acute Care Surg. 2013;74:339–43.
17. Pare' A. The apologie and treatise containing the voyages made into divers places with many of his writings upon surgery, translated out of Latine and compared with the French by Th. Johnson, Th. Cotes and R. Young, 1634. Reprinted by Falcon Educational Books, London, 1951.

18. Moyle J. Chirurgus Marinus or The Sea-Chirugion. 4th ed. London: E. Tracy; 1702.
19. Heister L. A general system of surgery, English translation. London: Innys; 1743.
20. Tripler CS, Blackman GC. Handbook for the military surgeon. Cincinnati: Robert Clarke & Co.; 1861.
21. The Medical and Surgical History of the War of the Rebellion. Part 1, Vol. 2 (Surgical History). US Army Surgeon Generals Office, Washington: Government Printing Office, 1870.
22. Holderman HH. Fracture and dislocation of the sternum. Ann Surg. 1928;88:252–9.
23. DeBakey M. The management of chest wounds: collective review. Int Abstract Surg. 1942;74:203–37.
24. Gray HK. War injuries of the chest. Staff Meetings of the Mayo Clinic 1942;17:566–72.
25. Leavitt DG. Non-union of three ribs. J Bone Joint Surg Am. 1942;24:932–6.
26. Elkin DC, Cooper FW. Thoracic injuries: review of cases. Surg Gyn Obstet. 1943;77:271–8.
27. Richardson DJ, Papper EM. Nerve-blocking therapy for fractured ribs. J Thorac Surg. 1947;16:432–7.
28. Hagen K. Multiple rib fractures treated with a Drinker respirator. J Bone Joint Surg. 1945;27:330–4.
29. Shefts L, Doud E. The management of thoracic and thoracoabdominal wounds in the forward areas in the Sicilian and Italian campaigns. J Thorac Surg. 1946;15:205–23.
30. Coleman FP, Coleman CL. Fracture of the ribs–a logical treatment. Surg Gyn Obstet. 1950;90:129–34.
31. Carter BN, Giuseffi J. Tracheotomy, a useful procedure in thoracic surgery, with particular reference to its employment in crushing injuries of the thorax. J Thorac Surg. 1951;21:495–505.
32. Valle A. Management of war wounds of the chest. J Thorac Surg. 1952;24:457–81.
33. Crutcher R, Nolen T. Multiple rib fracture with instability of chest wall. J Thorac Surg. 1956;32:15–21.
34. Bemelman M, Poeze M, Blokhuis TJ, Leenan LPH. Historic overview of treatment techniques for rib fractures and flail chest. Eur J Trauma Emerg Surg. 2010;36:407–15.
35. Constantinescu O. A new method of treating the flail chest wall. Am J Surg. 1965;109:604–10.
36. Avery E, Mörch ET, Benson D. Critically crushed chests: A new method of treatment with continuous mechanical hyperventilation to produce alkalotic apnea and internal pneumatic stabilization. J Thorac Surg. 1956;32:291–311.
37. Henry L. Case of stove-in chest injury. Br Med J. 1957;2:339.
38. Sillar W. The crushed chest: management of the flail anterior segment. J Bone Joint Surg. 1961;43B:738–45.
39. Le Roux BT, Stemmler P. Maintenance of chest wall stability: a further report. Thorax. 1971;26:424–8.
40. Judet R. Osteosynthese costale. Rev Chir Orthop. 1973;59(suppl):334.
41. Sanchez LJ. Osteosintesis costal median placa extraperiostica. Techn Rev Quirurgica Espanola. 1974;1:69.
42. Paris F, Tarazona V, Blasco E, et al. Surgical stabilization of traumatic flail chest. Thorax. 1975;30:521–7.
43. Menard A, Testart J, Philippe JM, Grise P. Treatment of flail chest with Judet's struts. J Thorac Cardiovasc Surg. 1983;86:300–5.
44. Moore B. Operative stabilization of nonpenetrating chest injuries. J Thorac Cardiovasc Surg. 1975;70:619–30.
45. Richardson JD, Grover FL, Trinkle JK. Early operative management of isolated sternal fractures. J Trauma. 1975;15:156–8.
46. Thomas A, Blaisdell W, Schlobohm R. Operative stabilization for flail chest after blunt trauma. J Thorac Cardiovasc Surg. 1978;75:793–801.
47. Meier P, Schupback P. Sur therapie des instabilen thorax bei rippenserienfrakturen. Schweiz Med Wschr. 1978;108:608–13.
48. Vecsei V, Frenzel I, Plenk H. A new rib plate for the stabilization of multiple rib fractures and thoracic wall fracture with paradoxical respiration. Hefte Unfallheilkd. 1979;138:279–82.

49. Labitzke R. Early thoracotomy and chest wall stabilization with elastic rib clamps. Zentralbl Chir. 1981;106:1351–9.
50. Hellberg K, de Vivie ER, Fuchs K, et al. Stabilization of flail chest by compression osteosynthesis–experimental and clinical results. Thorac Cardiovasc Surg. 1981;29:275–81.
51. Lee R, Bass S, Morris J, MacKenzie E. Three or more rib fractures as an indicator for transfer to a level 1 trauma center: a population-based study. J Trauma. 1990;30:689–94.
52. Garcia V, Gotschall C, Eichelberger M, Bowman L. Rib fractures in children: a marker of severe trauma. J Trauma. 1990;30:695–700.
53. Bulger E, Arneson M, Mock C, Jurkovich G. Rib fractures in the elderly. J Trauma. 2000;48:1040–7.
54. Engel C, Krieg J, Madey S, Long W, Bottlang M. Operative chest wall fixation with Osteosynthesis plates. J Trauma. 2005;58:181–6.
55. Haasler G. Open fixation of flail chest after blunt trauma. Ann Thorac Surg. 1990;49:993–5.
56. Gasparri MG, Tisol WB, Haasler GB. Rib stabilization: lessons learned. Eur J Trauma Emerg Surg. 2010;36:435–40.
57. Landercasper J, Cogbill TH, Lindesmith LA. Long-term disability after flail chest injury. J Trauma. 1984;24:410–4.
58. Mouton W, Lardinois D, Furrer M, Regli B, Ris H. Follow-up of patients with operative stabilization of a flail chest. Thorac Cardiovasc Surg. 1997;45:242–4.
59. Kerr-Valentic M, Arthur M, Mullins R, Pearson T, Mayberry C. Rib fracture pain and disability: can we do better? J Trauma. 2003;54:1058–64.
60. Ahmed Z, Mohyuddin Z. Management of flail chest injury: internal fixation versus endotracheal intubation and ventilation. J Thorac Cardiovasc Surg. 1995;110:1676–80.
61. Tanaka H, Yukioka T, Yamaguti Y, et al. Surgical stabilization or internal pneumatic stabilization? A randomized study of management of severe flail chest patients. J Trauma. 2002;52:727–32.
62. Voggenreiter G, Neudeck F, Aufmkolk M, et al. Operative chest wall stabilization in flail chest–outcomes of patients with and without pulmonary contusion. J Am Coll Surg. 1998;187:130–8.
63. Mayberry J, Ham B, Ellis T, Mullins R. A Phase IV clinical trial of the RibLoc Rib fracture repair system (poster). American Association for the Surgery of Trauma Annual Meeting, Pittsburgh, PA, 2009.
64. Sales R, Ellis T, Gillard J, et al. Biomechanical testing of a novel, minimally invasive rib fracture plating system. J Trauma. 2008;64:1270–4.
65. Bellezzo F, Hunt R, Provost P, Bain F, Kirker-Head C. Surgical repair of rib fractures in 14 neonatal foals: case selection, surgical technique and results. Equine Vet J. 2004;36:557–62.
66. Hoh D, Hoh B, Amar A, Wang M. Shape memory alloys: metallurgy, biocompatibility, and biomechanics for neurosurgical applications. Neurosurgery. 2009;64(5 suppl 2):199–214.
67. Jayle C, Allain G, Ingrand P, Laksiri L, et al. Flail chest in polytraumatized patients: surgical fixation using Stracos reduces ventilator time and hospital stay. Biomed Res Int. 2015;2015:624723. https://doi.org/10.1155/2015/624723.
68. Insertion of metal rib reinforcements to stabilise a flail chest wall Interventional procedures guidance [IPG361] National Institute for Health and Health Care Excellence. Published date: October 2010 www.nice.org.uk/guidance/ipg361/evidence, Accessed June 2017.
69. Bhatnagar A, Mayberry J, Nirula R. Rib fracture fixation for flail chest: what is the benefit? J Am Coll Surg. 2012;215:201–5.
70. Marasco S, Davies A, Cooper J, et al. Prospective randomized controlled trial of operative rib fixation in traumatic flail chest. J Am Coll Surg. 2013;216:924–32.
71. Doben A, Ericksson E, Denlinger C, et al. Surgical rib fixation for flail chest deformity improves liberation from mechanical ventilation. J Crit Care. 2014;29:139–43.
72. Pieracci F, Rodill M, Stovall R, et al. Surgical stabilization of severe rib fractures. J Trauma Acute Care Surg. 2015;78:883–7.

73. Tagawa T, Itoh S, Ide S, et al. Repair of intrathoracic visceral damage using video-assisted thoracoscopic surgery for blunt chest trauma and rib fixation at the site of mini-thoracotomy. Jpn J Thorac Cardiovasc Surg. 1998;46:121–6.
74. Sing D, Mostafa G, Matthews B, et al. Thoracoscopic resection of painful multiple rib fractures. J Trauma. 2002;52:391–2.
75. Mayberry J, Ham B, Ellis T. Thoracoscopic assisted rib fracture repair. Trauma Surgery Video Session, 92nd Annual Clinical Congress, American College of Surgeons, Chicago, IL, 2008.
76. Pieracci FM, et al. Completely thoracoscopic, intra-pleural reduction and fixation of severe rib fractures. Trauma Case Rep. 2015;1(5–8):39–43. https://doi.org/10.1016/j.tcr.2015.10.001.
77. Bemelman M, van Baal M, Yuan ZH, Leenen L. The role of minimally invasive plate osteosynthesis in rib fixation: a review. Korean J Thorac Cardiovasc Surg. 2016;49:1–8.

Chest Wall Anatomy

2

Noelle N. Saillant

The chest wall anatomy has evolved to efficiently support the mechanics of cardiopulmonary physiology. The skin and soft tissue overlie the musculature of the bone foundation creating the barrel-like structure of the chest wall. This architecture protects the thoracoabdominal viscera and allows for efficient volumetric changes in the chest dimension. The bony thorax, in conjunction with the musculature of the chest wall, moves in a pendelluft changing dimension with each breath to alter the intrathoracic pressures with inspiration and expiration. The inspiratory and expiratory pressure gradients assist the ventilation of the lungs, which are mechanically in series with the chest wall. A thorough understanding of the anatomy of the chest wall leads to appreciation of the physiology of respiration and gas exchange and is essential to the surgeon.

Surface Anatomy

On initial inspection of the chest, the anatomy is notable for important surface landmarks. The anatomical midline of the chest wall lies between the sternal notch and the xiphoid and continues distally through the linea alba between the rectus abdominis muscles. This central demarcation is important for gaining midline access for a median sternotomy [1] (Fig. 2.1).

As the sternum continues inferiorly, the angle formed by the junction of the manubrium and the body of the sternum, the angle of Louis, overlies the level of the aortic arch and the tracheal bifurcation [2].

As one proceeds toward the neck, the sternal and clavicular heads of the sternocleidomastoid mastoid muscles form Sedillot's triangle serving as reproducible

N. N. Saillant
Division of Trauma, Emergency Surgery and Surgical Critical Care, Department of Surgery, Massachusetts General Hospital, Harvard Medical School, Boston, MA, USA
e-mail: NSAILLANT@PARTNERS.ORG

M. de Moya, J. Mayberry (eds.), *Rib Fracture Management*,
https://doi.org/10.1007/978-3-319-91644-6_2

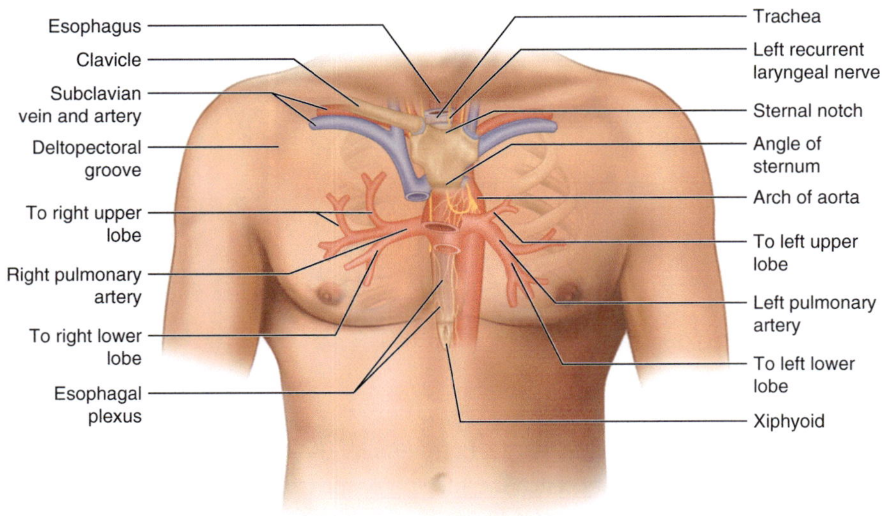

Fig. 2.1 Surface anatomy of the chest

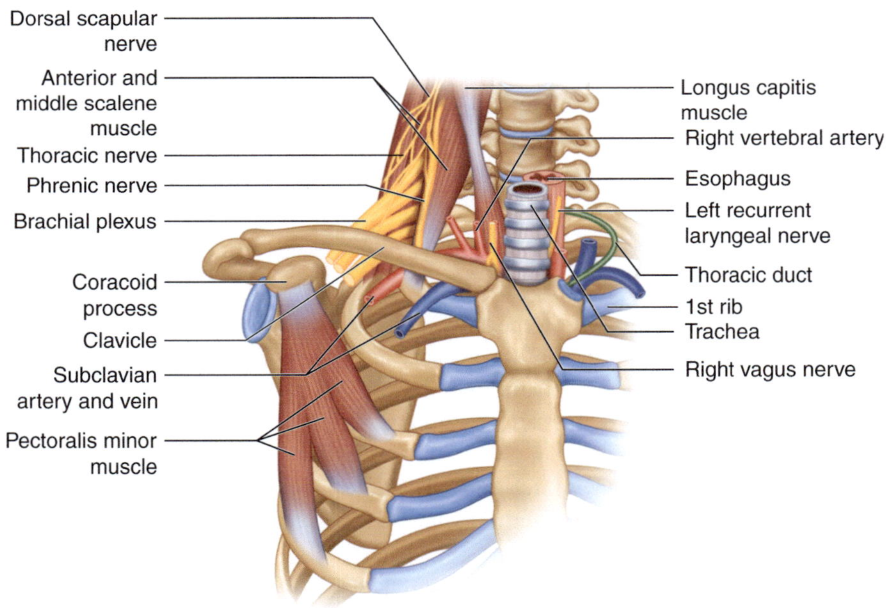

Fig. 2.2 The thoracic outlet

landmark for the cannulation of the internal jugular veins which lie posterior to the apex of the triangle [3]. Posterolateral to the sternocleidomastoid is the scalene muscle group. The scalene muscles originate on cervical vertebrae and insert on the first and second ribs to help assist in inspiration (Fig. 2.2).

Together with the costoclavicular space and the pectoralis minor, the scalenes form the thoracic outlet. Here, the subclavian vein and the phrenic nerves run along the anterior surface of the anterior scalene. On the left side of the chest, the thoracic duct courses along the left anterior scalene back to the junction of the subclavian and internal jugular veins to drain into the systemic circulation. One level deeper, the subclavian artery and the brachial plexus roots run between the anterior and middle scalene toward the upper extremity. Finally, the long thoracic and dorsal scapular nerves run between the middle and posterior scalene [4]. The presence of an anomalous rib, an enlarged C7 transverse process, or the presence of a fibrocartilaginous band may lead to compression of the neurovascular structures in the thoracic outlet. Known as thoracic outlet syndrome (TOS), the compression of the subclavian artery, vein, and/or brachial plexus can lead to embolic complications and cerebrovascular insufficiency, thrombosis, pain, and paresthesias, respectively [5].

Continuing laterally along the superior aspect of the chest, the clavicles protrude prominently to serve as a reproducible landmark for the cannulation of the subclavian vein just inferior to its middle third. The clavicular midpoint, in the second intercostal space, is one anatomic position for needle decompression of the chest for treatment of tension pneumothoraces, although current teaching is that the primary position is at the 5th intercostal space anterior axillary line.

As the chest spans toward the upper extremity, the deltopectoral groove aligns with the pathway of the axillary artery and vein as they pass from the first rib under the pectoralis minor. This vascular bundle and the cords of the brachial plexus are enclosed by the axillary fascia and attached to the clavipectoral fascia as the artery proceeds to the pyramidal shape space known as the axilla. The axilla is formed superiorly by the first rib, the distal clavicle, and the scapula and medially by the serratus anterior and underlying chest wall. The anterior border is formed by the pectoralis major and minor muscles and posteriorly by the teres major. It is upon crossing the teres major that the axillary vessels become the brachial vasculature. The lateral aspect of the axilla is determined by the intertubercular groove of the humerus.

The surface anatomy of the posterior chest is most notable for the spinous processes of the vertebral bodies along the axial spinal column and the prominent scapula spanning the first through seventh ribs.

Bony Thorax

Sternum

The sternum serves as the broad based anterior "shield" of the chest. The manubrium is the most superior aspect of the sternum. The superior lateral aspect of the manubrium is adjoined to the clavicle a via the sternoclavicular ligament. Laterally, the manubrium articulates with the cartilage of the first rib. The clavicle and the first rib are fused together at the costoclavicular ligament. Inferiorly the manubrium tethers to the sternal angle and the main body of the sternum (Fig. 2.3). This junction is marked laterally by the second rib, or inferiorly the sternum tapers to the xiphoid.

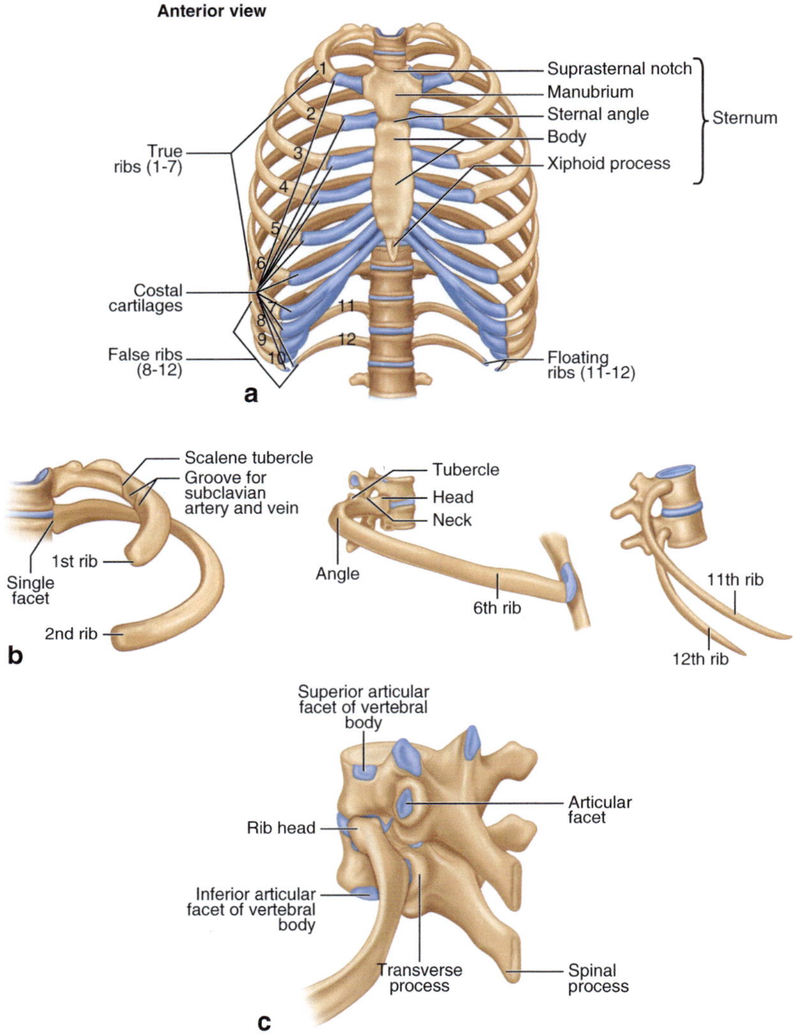

Fig. 2.3 The sternum and ribs

The Rib Cage

Twelve ribs create the hemithorax and provide the foundation for the muscular attachment of the chest wall. They provide the rigid support to protect the lungs and mediastinum. An inferior groove along the length of the rib houses the neurovascular bundle encompassing the intercostal artery, intercostal vein, and intercostal nerve unique to each rib space.

The first and second ribs serve as the apical cap of the chest and vary from the anatomy of ribs 3–10, the so-called typical ribs. The first rib is adjoined to the manubrium of

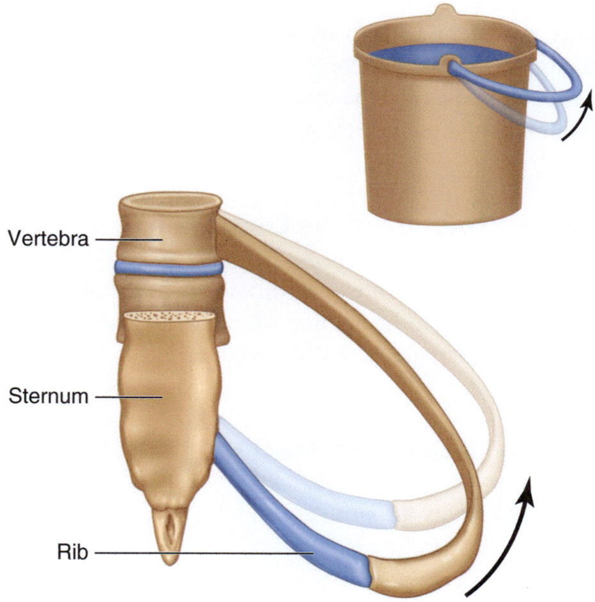

Fig. 2.4 Respiratory motion

the sternum. It is diminutive in size with a sharp curve to its articulation with the axial spine. The flatter shaft of the first rib is grooved by the passage of the subclavian vessels. The second rib articulates with the sternal manubrial junction and is larger than the first. Structurally, it is closely intertwined with the anterior scalene [1, 2].

Typical ribs (ribs 3–10) articulate with the transverse processes of the vertebral bodies via the most posterior aspect of the rib, called the head. The head gives way to the leanest part of the rib termed the "neck" that laterally becomes the articular facet and tubercle to which the costal cartilage attaches. The shaft of the rib provides the bow-like structure that creates the cylindrical shape of the chest wall. Finally, the typical rib is affixed anteriorly to its sternal articulation surface by the costal cartilage. The costal cartilage joins the ribs and sternum and marries the upward motion of the sternum to an anterior excursion of the ribs in a "bucket handle" movement. Ribs 1–7 directly articulate with the manubrium/sternum, whereas ribs 8–10 fuse anteriorly into the costal arch that fuses with the anterior cartilage of the costal margin of the 7th rib.

The most caudal ribs are the 11th and 12th ribs, collectively known as the floating ribs. They have only a posterior articulating surface on the head. The shafts of the 11th and 12th ribs taper to a cartilaginous endpoint that is not adjoined to the rest of the rib cage and is "floating" over the thoracoabdominal region (Fig. 2.4).

The Spine

The rib cage and the axial spine meet in two dimensions. The rib approaches the spine laterally at the vertebral body interspace. The head of a single rib meets with the articular surfaces of the vertebral body above and below the rib. The rib has a posterior lateral bend to allow its tubercle to associate with the transverse process of the more inferior vertebral body (i.e., the 5th rib head articulates with T4 and T5 vertebral bodies and the transverse process of T5). All the ribs have a robust, three-point articulation with the spine [1, 2].

The Musculature of the Chest Wall

The muscular structure of respiration is functionally separated by whether the muscle contributes to inspiration (through rib and sternal expansion and elevation) or expiration (though chest wall compression and depression of the thoracic cavity).

Primary Respiratory Muscles

The intercostals are primary muscles of respiration. They are comprised of three layers:

1. The external intercostal muscles which attach to the 1st–11th ribs and are inspiratory muscles
2. The internal intercostal muscles which insert inferiorly and posteriorly on the ribs and are muscles of expiration
3. The innermost intercostals, the subcostal muscles, and transversus thoracis which separate from the pleura via the endothoracic fascia

The diaphragm provides the largest contribution to the respiratory motion. As a primary muscle of inspiration, the diaphragm increases the chest wall dimension to draw air in. The dome-like curvature of the muscle contracts and descends toward the abdomen while elevating the attached ribs. The phrenic nerve is the primary innervation of the diaphragm. The blood supply is from the pericardiophrenic artery and musculophrenic arteries [1]. The superior lie of the diaphragm is at the level of the nipple line, or inframammary fold. In trauma this landmark divides the thoracic cavity from the abdomen. A breach in this line suggests the potential for chest, diaphragm, and abdominal visceral injury during the triage of penetrating trauma.

Accessory Muscles of Inspiration

The sternocleidomastoid muscles elevate the sternum, and the scalenes assist inspiration by lifting and elevating the superior rib couplet. Other accessory muscles of inspiration include the pectoralis major and minor, the serratus anterior, and the iliocostalis cervicis.

Accessory Muscles of Expiration

Serratus posterior superior aids in respiration by forcing respiration by lifting the superior ribs through its attachment from the nuchal ligament to the superior aspects of ribs 2–5. Its counterpart, the serratus posterior inferior, originates from the spinous processes of the thoracolumbar spine and contacts the posterior, inferior portion of ribs 9–12 and may help force air out of the chest in a pressured expiration (Fig. 2.5). Innervation of both muscles is from the intercostal arteries and nerves.

Latissimus dorsi encompasses a large span of the posterior chest wall. Latissimus dorsi originates from the ilium and sacrum with attachments to the thoracolumbar spinous process and ribs 8–12. The muscles inserts onto the humerus to adduct and internally rotate the arm. It is primarily supplied by the thoracodorsal artery and nerve.

The rectus abdominis, the external oblique, the internal obliques, and transversus abdominis muscles are considered part of the abdomen. They are, however, accessory muscles of respiration [2, 7–9].

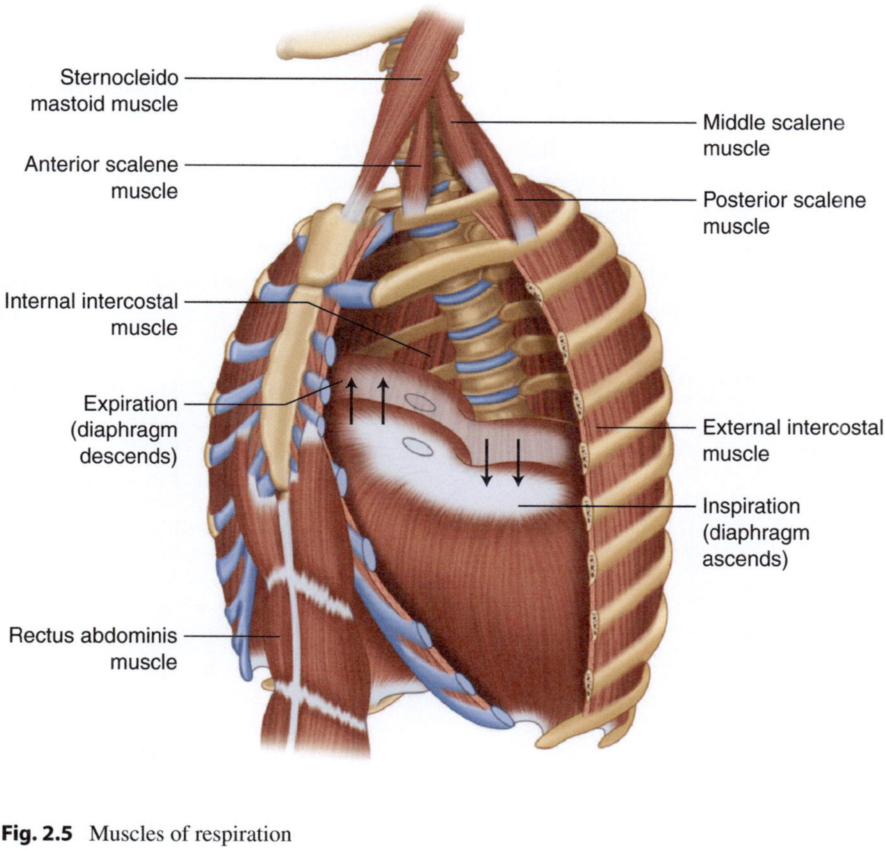

Fig. 2.5 Muscles of respiration

Nerves

Each layer of the intercostal muscle groups is supplied by the primary ventral rami of the 1st–11th spinal nerves which are called the intercostal nerves. The 1st–3rd intercostal nerves also contribute to the innervation of the upper extremity. The 4th–11th intercostal nerves supply the chest wall and intercostal muscles with inferior 7th–11th nerves contributing to the sensory innervation of the abdomen, diaphragm, pleura, and peritoneum. The 12th thoracic nerve is known separately as the subcostal nerve (Fig. 2.6). The perforating branches of the intercostal nerves innervate the soft tissue and skin of the chest. Additional nerve structures course through the chest wall and may be injured by trauma or surgery.

The long thoracic nerve originates from the anterior rami of the 5th, 6th, and sometimes 7th cervical nerves. It descends posterior to the brachial plexus and axillary vessels along the posterior axilla. It continues to the posterior surface of the latissimus dorsi and lies on the anterior surface of its target the serratus anterior. Injury to this nerve results in the clinical deformity known as the winged scapula [6].

The thoracodorsal nerve also runs in the posterior axillary wall. At its origin, it lies posterior to the artery; however with its descent down the chest, it becomes anterior to the thoracodorsal artery.

The intercostal brachial nerve is a commonly injured nerve in axillary surgery. It exits the second intercostal interspace at the midaxillary line and perforates the serratus anterior muscle to enter the subcutaneous tissue of the axilla. Damage to this structure leads to numbness of the axilla and medial arm.

The medial pectoral nerve runs medial and posterior to the axillary artery at its origin from the medial brachial plexus cord eventually having an anterior lie to the artery as it courses inferiorly on the chest. The lateral pectoral nerve runs anterior to the axillary artery with a small contribution to the medial pectoral nerve that wraps around the artery called the ansa pectoralis [1].

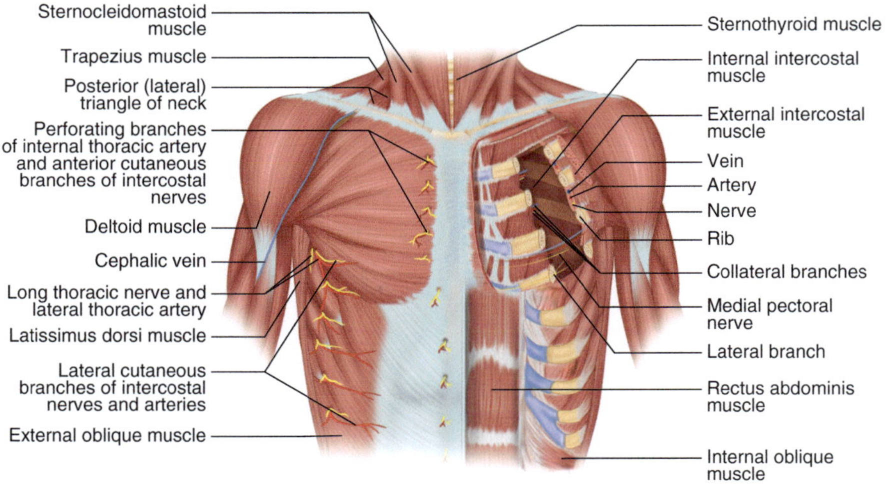

Fig. 2.6 Nerves of the anterior thoracic wall

Vasculature of the Chest Wall

The subclavian vessels give off the internal mammary vessels which run along either side of the sternum. These vessels subdivide into the intercostals arteries and veins which run laterally on the inferior groove of their respective ribs. Caudally the internal mammary vessels branch into the superior epigastric vessels and the musculophrenic artery and vein.

The superior thoracic artery is a division from the axillary artery that runs along the upper border of the pectoralis minor muscle to supply the upper intercostal spaces and the upper serratus anterior. The thoracoacromial trunk branches from the axillary artery into the pectoral, acromial, deltoid, and clavicular branches.

The lateral thoracic artery runs with the long thoracic nerve. It supplies blood to the pectoralis, axilla, and subscapularis muscles. The thoracodorsal artery runs with its respective nerve bundle and supplies latissimus dorsi and the inferior serratus anterior.

[7–10].

In conclusion, the anatomy of the chest wall marries form and function. Appreciating the details of its design enables the surgical hand and mind in the approach to chest wall surgery and the treatment of thoracic trauma.

References

1. Netter FM. The ciba collection of medical illustrations. In: Respiratory system, vol. 7. Ardsley, NY: Ciba-Geigy Corporation; 1979.
2. Miller JI, BRyant A, Deskauriers J. Anatomy and the physiology of the chest wall ad sternum with surgical implications. Pearson's Thoracic and Esophageal Surgery, chapter 98, 1197–1208 Churchill Livingstone, 2008.
3. Mp B, Heller SF, Rivera M. Anatomic considerations for central venous cannulation. Risk Manag Health Policy. 2011;4:27–39. Published online 2011 Apr 13
4. Rusnak-Smith S, Moffat M, Rosen E. Anatomical variations in the scalene triangle: dissection of 10 cadavers. J Orthop Sports Phys Ther. 2001;31(2):70–80.
5. Hussain MA, Aljabri B, Al-Omran M. Vascular thoracic outlet syndrome. Semin Thorac Cardiovasc Surg. 2016;28(1):151–7.
6. Gooding BW, Geoghan JM, Wallace WA, Manning PA. Scapular winging. Shoulder Elbow. 2014;6(1):4–11.
7. Naidu BV, Rajesk PB. Relevant surgical anatomy of the chest wall. Thorac Surg Clin. 2010;20:453–63.
8. Graeber G, Nazim M. The anatomy of the ribs and the sternum and their relationship to chest wall structure and function. Thorac Surg Clin. 2007;17:473–89.
9. Carrier G, Frechette E, Ugalde P, Desauliers J. Correlative anatomy for the sternum, ribs, costovertebral angle, chest wall, muscles and intercostal spaces, thoracic outlet. Thorac Surg Clin. 2007;17:521–8.
10. Westaby S, Odell JA. Cardiothoracic trauma. In: The pathophysiology of chest trauma. New York, NY: Oxford University Press. INC; 1999.

Pathophysiology of Rib Fractures and Lung Contusion

3

Jeffrey J. Skubic, Barbara U. Okafor, and Deepika Nehra

Anatomy and Physiology

The bony thoracic cage is a perfect example of how structure and function are intimately related. The chest wall is such that the bony structures (ribs, sternum, and vertebrae) work in conjunction with the muscular structures to allow for respiration to occur. A detailed understanding of chest wall anatomy will help in understanding how traumatic injuries affecting the structure of the thorax can affect respiration.

Anatomy of the Chest Wall

Ribs

The 12 ribs of the thorax are divided into *true* and *false ribs*. The first seven ribs are true ribs as they form a complete loop between the sternum and vertebrae. The lower five ribs are considered false ribs as they do not fully reach the sternum anteriorly. Of these lower five ribs, ribs 7 through 10 connect to the cartilage of the rib above them and therefore connect to the sternum indirectly. Ribs 11 and 12 are considered *floating ribs*.

Ribs 3 through 10 each have a head, neck, and body and are considered *typical ribs*. Typical ribs (with the exception of rib 10) have two articulations on the head, one connected to the vertebra of the same level and the other connected to the vertebra one level superior. At the junction of the head and neck, there is a tubercle that articulates with the transverse process of the associated vertebra. After the tubercle, the body of the rib continues and at the costal angle starts to wrap around anteriorly heading toward the sternum. The costal angle also marks the lateral extent of the

J. J. Skubic · B. U. Okafor · D. Nehra (✉)
Division of Trauma, Burns and Surgical Critical Care, Department of Surgery, Brigham and Women's Hospital, Harvard Medical School, Boston, MA, USA
e-mail: dnehra@bwh.harvard.edu; JSKUBIC@PARTNERS.ORG

M. de Moya, J. Mayberry (eds.), *Rib Fracture Management*,
https://doi.org/10.1007/978-3-319-91644-6_3

attachment of the deep back muscles. Anteriorly ribs 1 through 6 will connect to the sternum, while ribs 7 through 10 will connect indirectly through their fused cartilages. The remaining ribs (ribs 1–2 and 11–12) are considered *atypical ribs*. The first rib is very broad and sharply curved. Its head articulates only with the first thoracic vertebra. The second rib is thinner and slightly less curved, and its head articulates with both the first and second thoracic vertebrae. Ribs 11 and 12 have single articulations with their associated vertebrae and are short, have no neck, and are free-floating.

Sternum

The sternum consists of the manubrium, the body, and the xiphoid process. The clavicles as well as the first and second ribs articulate with the manubrium. The manubrium then joins with the sternal body at the sternal angle of Louis; this joint may become ossified later in life limiting motion. Ribs 2 through 7 articulate laterally with the sternal body through costal cartilage. Inferiorly, the sternal body gives rise to the cartilaginous xiphoid process.

Muscles

There are 17 muscles of the chest wall; a complete discussion of these muscles including those used mostly for attachment to appendages is beyond the scope of this chapter; however, muscles involved in the process of respiration will be discussed.

The *true muscles* of the thoracic wall are the serratus posterior, levatores costarum, intercostal, subcostal, and transversus thoracis muscles. The serratus posterior originates at the spinous processes of the vertebrae and inserts onto the superior border of the upper ribs. On the lower ribs, the insertion is on the inferior border of the ribs. Where once these muscles were thought to assist in inspiration, more recent research has suggested that their primary function is actually rib proprioception. Levatores costarum run from the transverse processes of seventh through the eleventh thoracic vertebrae to the ribs specifically between the tubercle and the angle of the ribs. These muscles elevate the ribs during inspiration. The external, internal, and innermost intercostal muscles are located between the ribs and function primarily to hold the chest wall rigid during respiration but can also contribute to the actual motion of breathing such as helping to elevate the ribs during forced respiration. The subcostal muscles run from the inner surface of a rib to the inner surface of a rib one or two spaces below it. The transversus thoracis runs from the posterior surface of the sternum to the inner surfaces of costal cartilages two through six. It is continuous with the transversus abdominis muscle in the abdomen. It contributes weakly to expiration and has some proprioceptive abilities.

The diaphragm is a domed sheet of skeletal muscle that separates the thorax from the abdomen. It is innervated by the phrenic nerve and serves as the chief muscle of respiration.

Accessory muscles of respiration include the serratus anterior, pectoralis, and scalene muscles and may assist in elevating the ribs at times.

Respiratory Physiology

During inhalation, the diaphragm contracts and moves in the inferior direction creating negative pressure in the thoracic cavity which draws air into the lungs. Simultaneously, the accessory muscles contract causing the ribs to move anteriorly, superiorly, and laterally which increases the volume of the thoracic cavity, further increasing negative pressure in the thorax. In this way, the chest wall helps to prevent paradoxical motion created by the negative pressure from diaphragmatic contraction. During exhalation, the diaphragm and accessory muscles relax decreasing intrathoracic volume and increasing pressure which allows the elastic recoil of the lungs to expel air. The sternal manubrium is fixed and does not move much with respiration. The upper ribs and body of the sternum have an anterior-posterior movement during breathing referred to as pump-handle motion. The lower ribs lift up and out laterally with breathing which is referred to as bucket-handle motion.

Pathophysiology

Rib Fractures

Rib fractures are the most common bony fracture among blunt trauma patients, occurring in about 10–40% of all blunt trauma patients [1]. Most rib fractures are a result of motor vehicle collisions, pedestrian struck, and falls [2]. The incidence of rib fractures in blunt trauma patients increases with age, from 25% in children (≤18 years of age) to 50% in adults (19–64 years of age) and 65% in elderly patients (≥65 years of age) [1]. In several studies, number of rib fractures has been directly correlated with mortality even from non-pulmonary causes. Having greater than six rib fractures has been shown to significantly increase mortality for all age groups [1, 2].

Rib fractures can disrupt the very specific structure of the thoracic cage and therefore interfere with the function of respiration. Rib fractures are also a marker of injury severity. Not only can the rib fractures themselves cause pain with respiration leading to splinting, inability to clear secretions, and pneumonia, but rib fractures can also be a sign of other underlying injuries. Fractures of the first rib are significant because of the high kinetic energy necessary to fracture these ribs. While an isolated, non-displaced first rib fracture is only associated with a 3% vascular injury rate, any other findings such as a concomitant head, thoracic, abdominal, or long bone injury increase this risk to 24% [3]. If the first rib is displaced posteriorly, the fracture involves the subclavian sulcus, there is any evidence of brachial plexus injury, or there are abnormal findings on chest roentgenogram (widened mediastinum, apical capping, hemothorax, tracheal deviation, left main stem bronchus depression, widened paratracheal stripe, loss of the aortopulmonary window, or an abnormal aortic contour), a subclavian artery or aortic injury needs to be ruled out. Second rib fractures are also associated with underlying neurovascular injury and should be treated similarly [4]. Fractures of ribs four through nine may be associated with underlying injuries to the heart, lungs, or bronchi. Although associated abdominal solid organ injury has been classically associated with fractures of ribs 10 through 12, a recent study demonstrates that patients with fractures involving ribs 5 through 12 are at risk for solid organ injury in the abdomen [5]. Furthermore, rib fractures have been associated not only with thoracic injuries but extra-thoracic injuries as well.

Flail Chest

A radiographic fail chest occurs when three or more contiguous ribs are broken in two or more places. As described above, as the diaphragm contracts during inhalation, the chest wall also expands due to contraction of the accessory muscles. The contraction of these accessory muscles allows the chest wall to resist the negative inspiratory pressure that results from contraction of the diaphragm. A flail chest creates the potential for a segment of that bony cage to have paradoxical movement with respiration. This type of paradoxical movement is referred to as a *clinical flail* and is most commonly observed if the involved ribs are fractured anterolaterally. Such paradoxical movement of this section of the chest wall can result in significant respiratory embarrassment. A recent retrospective National Trauma Data Bank (NTDB) study of blunt trauma patients admitted with a diagnosis of flail chest reviewed approximately 3500 cases. As expected, most were associated with motor vehicle collisions or falls. Just over half of the patients developed a pulmonary contusion, which is a tribute to how much kinetic energy the chest must absorb to create a flail chest. Additionally, 80% required admission to an intensive care unit, with mechanical ventilation being required in nearly 60%. The rates of pneumonia and acute respiratory distress syndrome (ARDS) were 21% and 14%, respectively, which is certainly at least partially a result of the need for mechanical ventilation. The mortality rate was 11% overall, which increased to 40% in those with a concurrent head injury [6]. Radiographic flail segments should alert the physician to a significant energy blunt force trauma, the potential for underlying associated injuries, and pulmonary contusions. A clinical flail is especially concerning for the potential for respiratory embarrassment.

Pulmonary Contusion

Lung or pulmonary contusions were first investigated during the wars of the twentieth century starting with World War I, where soldiers close to blast sites were found expired with little to no signs of external trauma. After nearly 4 years of these experiences and postmortem examinations, combat surgeons recognized the cause of these deaths as pulmonary contusions [7]. In the 1970s, animal studies on dogs described the pathophysiology as "blood and plasma" filling the alveoli causing a reduction in lung compliance and an increase in shunting [8]. Today, pulmonary contusions are recognized as the most common injury from blunt thoracic trauma. Half of patients with a flail chest will develop an associated pulmonary contusion. Typically pulmonary contusions "bloom" in about 3 days and resolve within a week; however, animal models have shown changes in the lung such as alveolar hemorrhage, atelectasis, and lung consolidation in as little as 24 h after blunt injury [7]. More recent porcine and murine studies have shown delayed pathologic changes in the lung contralateral to the injury as well as circulating cytokine markers indicating a systemic inflammatory and also immunologic response to a unilateral lung injury [9].

The effects of a pulmonary contusion are multifold and complex and can be divided into three categories: those that occur in the injured lung, those that occur in the injured and uninjured lung, and those that occur systemically. At the site of injury, the alveoli are hemorrhagic, and reduced compliance contributing to shunting and hypoxemia, increased pulmonary vascular resistance, and overall decreased pulmonary blood flow is present. In the injured as well as uninjured parts of the lung, including the contralateral lung, the alveolar septa become thickened, the parenchymal tissues become vacuolized, and there exist a delayed capillary leak and increased neutrophil migration within the parenchyma. The CT findings associated with a pulmonary contusion after a high blunt force injury to the chest are seen in Fig. 3.1. Systemically, there is increased terminal complement complex, tumor necrosis factor, and interleukin-6 with decreases in complement, peritoneal, and splenic macrophages as well as splenocytes [7].

Pulmonary contusions can impair myocardial performance even in the absence of a cardiac contusion. A porcine study has shown depressed myocardial contractility and increased right ventricular afterload in the presence of an isolated pulmonary contusion [10]. The authors recommended measuring central venous pressure (CVP) to estimate preload in the presence of pulmonary contusions as they demonstrated pulmonary capillary wedge pressure (PCWP) to underestimate

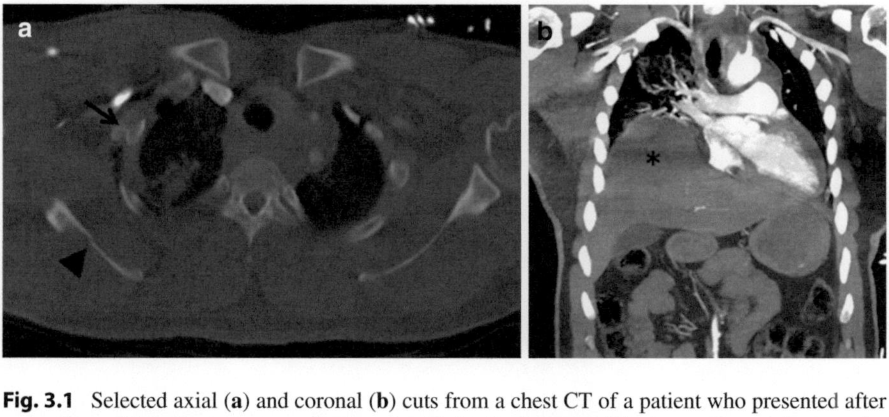

Fig. 3.1 Selected axial (**a**) and coronal (**b**) cuts from a chest CT of a patient who presented after a significant blunt force trauma. Seen are multiple rib fractures (*black arrow*), a right scapular fracture (*black arrowhead*) and associated pulmonary contusion, subcutaneous emphysema, and a right diaphragmatic rupture with the liver herniated into the right chest

left ventricular end diastolic pressure in their lab model. They showed CVP to be a better estimate of preload even with the pleural space being violated (a chest tube was inserted immediately after creation of the contusion.)

Pneumothorax

Greek physicians first described the pneumothorax as early as the fifth century B.C., but French physician, Itard, first coined the term in 1803. Pneumothoraces are estimated to occur in roughly 15% of blunt trauma patients, but the incidence has been reported to be as high as 80% in those with two or more broken ribs [11].

A pneumothorax refers to the presence of gas within the pleural space, i.e., between the visceral and parietal pleura. A pneumothorax associated with rib fractures is seen in a chest X-ray in Fig. 3.2 and chest CT scan in Fig. 3.3. Gas within the pleural space indicates either an open communication between the outside atmosphere and the pleural space (sucking chest wound) or a communication between air-containing intrathoracic structures, such as the alveoli or bronchi, and the pleural space. A rib fracture can lacerate the lung allowing open communication between the alveoli and the pleural space. It has been theorized that sudden compression of the chest can cause rupture of an alveolus without a laceration leading to a pneumomediastinum, also known as the Macklin effect. Once communication between an air-containing structure and the pleural space exists, air will flow into the pleural space until either the communication is closed or the pressure equilibrates.

In the case where pressure rises in the pleural space until mediastinal structures are shifted away from the side of the pneumothorax leading to compression of cardiac

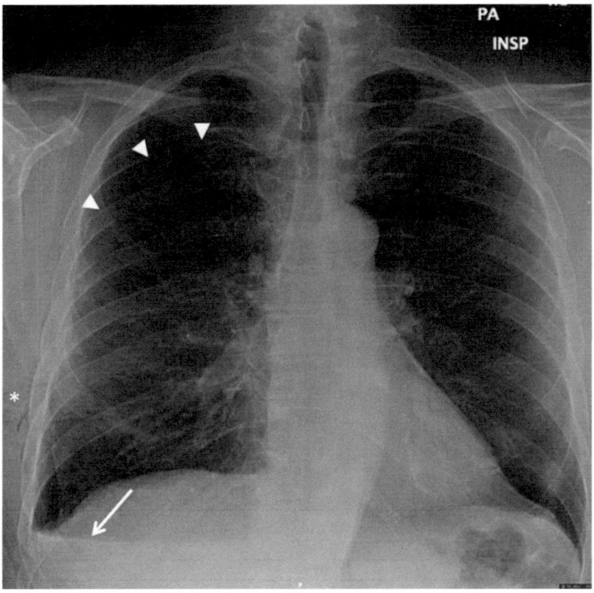

Fig. 3.2 Chest X-ray of a patient with a hemopneumothorax with a visible pneumothorax (*white arrowheads mark lung edge*) and an air-fluid level (*white arrow*). This patient also had multiple right-sided rib fractures that are poorly visualized on this X-ray and were better seen on a chest CT scan

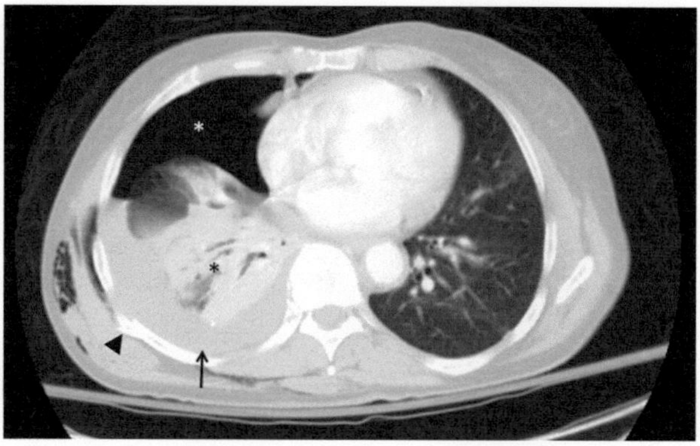

Fig. 3.3 Selected axial image from a chest CT of a patient involved in a blunt trauma resulting in multiple right-sided rib fractures (*black arrowhead*) in addition to a pneumothorax (*white star*), a hemothorax (*black arrow*), and an underlying pulmonary contusion (black star). Also seen is significant subcutaneous emphysema in the soft tissues of the right chest wall

inflow and cardiovascular compromise, this is called a *tension pneumothorax*. This is caused by a one-way valve that allows air to enter the pleural space but not exit. A tension pneumothorax is a clinical diagnosis not a radiographic one, with decreased breath sounds on the affected side, distended neck veins, a contralaterally deviated trachea, and hypotension. Subcutaneous emphysema should also hint at an underlying airway injury.

Hemothorax

A significant number of patients with chest trauma have hemothoraces [12]. Any patient with a pneumothorax should be suspected to have a hemothorax as many of these are actually hemopneumothoraces. CXR findings consistent with a traumatic hemopneumothorax are seen in Fig. 3.2, and chest CT findings are seen in Fig. 3.3. Rib fractures can commonly cause a hemothorax by lacerating any source of bleeding within the chest such as an intercostal vessel, intramuscular vessel, pulmonary vessel, or lung parenchyma. The majority of hemothoraces are diagnosed at admission, but up to one-third have been reported as a delayed hemothorax, which may present with respiratory symptoms or pain several days after the initial trauma.

Special Considerations

Geriatric Patients

Falls and motor vehicle collisions are the most common causes of rib fractures in the elderly (≥65 years of age) with some studies reporting falls as contributing more and more to rib fractures in this population [13]. As mentioned earlier, elderly blunt trauma patients have a higher incidence of rib fractures as compared to younger cohorts of patients, but more importantly, these patients also have worse outcomes even after adjusting for comorbidities and injury severity [14, 15]. The higher incidence of rib fractures among elderly patients is a result of lower bone density and ossified costal cartilages causing the chest wall to be more brittle and less compliant [2]. Additionally the likelihood of associated thoracic injuries is high with nearly 70% of elderly patients with rib fractures having at least one of the following: pulmonary contusion, infection, need for mechanical ventilation, hemothorax, or pneumothorax [16].

Pediatric Patients

Only 5% of pediatric traumatic patients have thoracic injuries; however, the mortality associated with thoracic injuries in this age group is 20% [17]. This mortality increases further if rib fractures are present. Motor vehicles are a common cause of thoracic trauma in pediatric patients, with a significant proportion being the result of the child being struck on a pedestrian by a motor vehicle [17]. Ribs in pediatric patients are more compliant and flexible and less likely to fracture as a result of blunt force trauma. This means that pediatric patients with rib fractures have often times sustained an extremely high-energy impact to the chest. Associated thoracic and extra-thoracic injuries are very common [2, 17]. Pulmonary contusions in pediatric patients with rib fractures tend to be quite severe and diffuse. Additionally, in the pediatric blunt trauma patient, underlying visceral injuries should be suspected even without rib fractures; however, the presence of even a single rib fracture should be concerning for the potential for severe underlying visceral injury due to the amount of kinetic energy needed to overcome the compliance of the flexible pediatric chest wall. The mortality for pediatric patients with rib fractures also increases linearly with each additional rib fracture without an inflection [18].

References

1. Sharma OP, Oswanski MF, Jolly S, Lauer SK, Dressel R, Stombaugh HA. Perils of rib fractures. Am Surg. 2008;74(4):310–4.
2. Majercik S, Pieracci FM. Chest wall trauma. Thorac Surg Clin. 2017;27(2):113–21.
3. Gupta A, Jamshidi M, Rubin JR. Traumatic first rib fracture: is angiography necessary? A review of 730 cases. Cardiovasc Surg. 1997;5(1):48–53.
4. Livoni JP, Barcia TC. Fracture of the first and second rib: incidence of vascular injury relative to type of fracture. Radiology. 1982;145(1):31–3.
5. Rostas JW, Lively TB, Brevard SB, Simmons JD, Frotan MA, Gonzalez RP. Rib fractures and their association with solid organ injury: higher rib fractures have greater significance for solid organ injury screening. Am J Surg. 2017;213(4):791–7.
6. Dehghan N, de Mestral C, McKee MD, Schemitsch EH, Nathens A. Flail chest injuries: a review of outcomes and treatment practices from the national trauma data bank. J Trauma Acute Care Surg. 2014;76(2):462–8.

7. Simon B, Ebert J, Bokhari F, et al. Management of pulmonary contusion and flail chest: an eastern Association for the Surgery of trauma practice management guideline. J Trauma Acute Care Surg. 2012;73(5 Suppl 4):S351–61.
8. Oppenheimer L, Craven KD, Forkert L, Wood LD. Pathophysiology of pulmonary contusion in dogs. J Appl Physiol Respir Environ Exerc Physiol. 1979;47(4):718–28.
9. Davis KA, Fabian TC, Croce MA, Proctor KG. Prostanoids: early mediators in the secondary injury that develops after unilateral pulmonary contusion. J Trauma. 1999;46(5):824–31. discussion 831-822
10. Moomey CB Jr, Fabian TC, Croce MA, Melton SM, Proctor KG. Determinants of myocardial performance after blunt chest trauma. J Trauma. 1998;45(6):988–96.
11. Liman ST, Kuzucu A, Tastepe AI, Ulasan GN, Topcu S. Chest injury due to blunt trauma. Eur J Cardiothorac Surg. 2003;23(3):374–8.
12. MacLeod JB, Ustin JS, Kim JT, Lewis F, Rozycki GS, Feliciano DV. The epidemiology of traumatic hemothorax in a level I trauma center: case for early video-assisted thoracoscopic surgery. Eur J Trauma Emerg Surg. 2010;36(3):240–6.
13. Earl-Royal E, Shofer F, Ruggieri D, Frasso R, Holena D. Variation of blunt traumatic injury with age in older adults: statewide analysis 2011–14. West J Emerg Med. 2016;17(6):702–8.
14. Shulzhenko NO, Zens TJ, Beems MV, et al. Number of rib fractures thresholds independently predict worse outcomes in older patients with blunt trauma. Surgery. 2017;161(4):1083–9.
15. Flagel BT, Luchette FA, Reed RL, et al. Half-a-dozen ribs: the breakpoint for mortality. Surgery. 2005;138(4):717–23. discussion 723-715
16. Sirmali M, Turut H, Topcu S, et al. A comprehensive analysis of traumatic rib fractures: morbidity, mortality and management. Eur J Cardiothorac Surg. 2003;24(1):133–8.
17. Skinner DL, den Hollander D, Laing GL, Rodseth RN, Muckart DJ. Severe blunt thoracic trauma: differences between adults and children in a level I trauma Centre. S Afr Med J. 2015;105(1):47–51.
18. Rosenberg G, Bryant AK, Davis KA, Schuster KM. No breakpoint for mortality in pediatric rib fractures. J Trauma Acute Care Surg. 2016;80(3):427–32.

4 Scoring Systems of Blunt Thoracic Trauma and Rib Fractures

Fredric M. Pieracci

Scoring systems for injury patterns represent a common language through which providers may communicate with each other and their patients, prognosticate, select therapies, educate, and conduct research. The purpose of a scoring system is to organize in a graded fashion the myriad of variability within a set of specific injuries in order to categorize patients based upon their risk of developing complications. Novel therapies can then be studied within these groups with the goal of identifying the optimal target patient population.

Some of the specific challenges of developing a chest wall injury scoring system have been the lack of a universal nomenclature to describe these injuries, overreliance on radiographic parameters, exclusion of long-term quality of life outcomes, and lack of validation of currently available systems among large, diverse patient populations. Despite these limitations, several validated scoring systems are available to guide the clinician in early risk stratification and resource allocation.

General Properties of Scoring Systems

The ideal diagnostic test predicts an outcome perfectly; each time the test is positive, the outcome is present, and each time the test is negative, the outcome is absent. This concept is quantified in the four basic test properties; sensitivity, specificity, positive predictive value, and negative predictive value (Table 4.1). Sensitivity is defined as the portion of patients with the outcome that test positive; it is therefore also known as the true positive rate. Highly sensitive tests minimize false negatives and are therefore useful for ruling in a condition. Specificity is defined as the portion of patients without the outcome that test negative; it is therefore also known as the

F. M. Pieracci
Department of Surgery, Denver Health Medical Center, University of Colorado School of Medicine, Denver Health Medical Center, Denver, CO, USA
e-mail: Fredric.pieracci@dhha.org

M. de Moya, J. Mayberry (eds.), *Rib Fracture Management*,
https://doi.org/10.1007/978-3-319-91644-6_4

Table 4.1 2 × 2 table illustrating the four basic test parameters

	Outcome is positive	Outcome is negative
Test is positive	True positive (TP)	False positive (FP)
Test is negative	False negative (FN)	True negative (TN)

Sensitivity = TP/(TP + FN); specificity = TN/(TN + FP); positive predictive value = TP/(TP + FP); negative predictive value = TN/(TN + FP)

true negative rate. Highly specific tests minimize false positives and are therefore useful for ruling out a condition. Neither sensitivity nor specificity is dependent upon the prevalence of the outcome. Values >80% are generally considered to be favorable. The main limitation of both sensitivity and specificity is that the outcome status of the patient must be known in order to apply the test clinically.

A more practical predictive parameter is the proportion of patients that test positive who go on to develop the outcome. This ratio is the positive predictive value (PPV). Its mirror, the negative predictive value (NPV), represents the proportion of patients that test negative who do not develop the outcome. Consider the hypothetical relationship between incentive spirometry (IS) <1 l and pneumonia among rib fracture patients. If this test were 85% sensitive, it would mean that 85% of patients who developed pneumonia had an admission IS <1 l. This information is of marginal clinical utility since it is unknown on admission who will develop the outcome of pneumonia. By contrast, a PPV of 85% means that 85% of patients with an admission IS <1 l will develop pneumonia. This information is far more practical for prognostic purposes. Importantly, the PPV and NPV are dependent upon the prevalence of the underlying outcome.

Prediction models may be either categorical or continuous. Furthermore, categorical models may be either binary or multilevel. The aforementioned example of an admission IS <1 l is binary; it is either present or absent. The use of the actual IS value would constitute a continuous variable. Finally, operationalizing the IS as poor, adequate, or good would be an example of a three-level, ordinal, categorical variable.

Most predictive models are either ordinal categorical or continuous. Such models therefore consist of multiple thresholds for predicting the outcome of interest. Consider the example of using age as a continuous variable to predict the need for tracheostomy among rib fracture patients. Patients may be grouped using an infinite number of age thresholds (e.g., age >20 years, age >21 year, etc.). Calculation of sensitivity and 1—specificity for the outcome of tracheostomy at each of these thresholds generates a curved line, known as the receiver operator characteristic (ROC) curve, which provides both a visual and mathematical representation of a test's predictive ability (Fig. 4.1). The quotient obtained by dividing the sensitivity by 1—specificity is known as the likelihood ratio (LR). A likelihood ratio close to one indicates that the test adds little information to the clinical picture.

The area under the ROC curve (AUC) is the standard statistical expression of a test's overall predictive ability. The AUC ranges from 0.5 to 1.0, with 0.5 being no better than chance and 1.0 being perfect prediction. In general, a test with an ROC

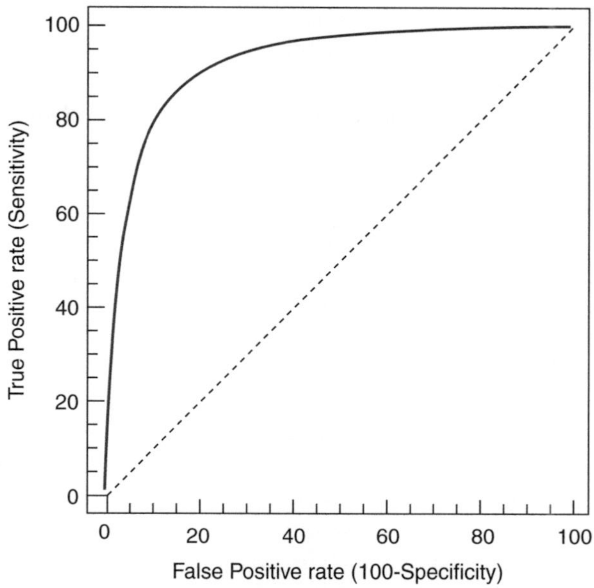

Fig. 4.1 Generic receiver operator characteristic curve. The dotted line represents a test with no predictive ability (area under the curve = 0.50)

AUC of 0.7–0.8 is considered moderate predictive ability, 0.8–0.9 good predictive ability, and 0.9–0.99 excellent predictive ability. Finally, the point of maximal distance between the ROC curve and the chance line is known as Youden's J and returns the most appropriate threshold for the diagnostic test. It is calculated as sensitivity + specificity – 1.

Ideal Characteristics of a Chest Wall Scoring System

Designing a high-fidelity chest wall injury scoring system involves careful consideration of both predictor and outcome variables. Predictor variables that are selected for inclusion into models must be evaluated not only for association with the outcome but also overall prevalence, ease of abstraction, objectivity, and interobserver reliability. Consider the variable "degree of fracture displacement." Although this variable may ultimately be important in predicting adverse outcomes, as written it is far too subjective. How would the degree of displacement be measured, by whom, and when? Would all fracture locations on all ribs be weighted evenly? A more specific variable might be "three or more fractures with bicortical displacement." As a second example, although open rib fractures are highly associated with both morbidity and mortality, they occur too rarely so as to provide meaningful contribution to a universal scoring system. In practice, there exists a trade-off between ease of abstraction and predictive ability. The total number of rib fractures is a common, easily abstracted parameter with low inter-rater variability. However, it is overly simplistic, failing to account for the difference in outcome by fracture location on the rib, degree of displacement, and presence or absence of a flail segment.

Most importantly, it does not account for the clinical phenotype of the injury, which varies widely among patients.

Selection of outcome variables also requires careful consideration. Although mortality is a grave complication, it is relatively infrequent among chest wall-injured patients. For example, in our recent multicenter analysis of timing of surgical stabilization of rib fractures, mortality occurred in less than 1 in 500 patients [1]. Furthermore, the use of a crude outcome such as mortality may confer a false sense of success among patients who survive. Consider the example of a patient who survives their chest wall injury but is permanently disabled, on a ventilator, and unable to work or enjoy any pre-injury activities. One major challenge in the care of rib fracture patients moving forward is the incorporation of long-term quality of life outcomes within scoring systems.

The final quality for consideration in the design of a scoring system is a statistical one. Most scoring systems used in trauma patients are linear, giving equal weight to each predictive variable. One example of such a system is the Blunt Pulmonary Contusion 18 Score [2], in which the same increment exists moving between no contusion (0 points) and a mild contusion (1 point) as does for moving from a moderate contusion (2 points) to a severe contusion (3 points). In reality, it is likely that these differences are not related linearly. Consideration of both nonlinear and weighed modeling is imperative to designing a high-fidelity system. Collaboration with biostatisticians when possible should be part of any model development.

One overarching limitation to the development of a universal grading system for chest wall injuries has been the lack of a uniform classification system. There remain large discrepancies in the interpretation of several commonly employed terms, such as flail chest, fracture displacement, and lateral rib fractures. The creation of a standardized nomenclature is currently the subject of a joint collaboration between the American Association for the Surgery of Trauma, the Chest Wall Injury Society, and the AO Thoracic Expert Group.

Current Chest Wall Injury Scoring Systems

Several single parameters have been used to predict outcomes of chest-injured patients separate from incorporation into a formal scoring system. These include the overall number of rib fractures [3, 4], number of displaced rib fractures [5], incentive spirometry [6], vital capacity [7], age [8], and degree of pulmonary contusion [2]. The following discussion will be limited to validated scoring systems that include at least two variables. Such scores are summarized according to the parameters that they include in Table 4.2.

Chest Wall Organ Injury Scale

The Chest Wall Organ Injury Scale (Table 4.3) is the oldest and most commonly used classification system for chest wall injuries, which is endorsed by the American

Table 4.2 Comparison of current chest wall injury scoring systems

	OIS chest	RFS	CTS	RibScore
Rib fracture variables				
Number of ribs fractured	•	•	•	•
Flail chest	•			•
Bilateral fractures	•	•	•	•
Degree of displacement				•
Fracture location				•
First rib fractured				•
Non-rib fracture variables				
Age	•	•	•	
Pulmonary contusion	•		•	
Clavicle/scapula/sternal fractures	•			

From Chapman BC, et al. RibScore: A novel radiographic score based on fracture pattern that predicts pneumonia, respiratory failure, and tracheostomy. J Trauma Acute Care Surg. 2016;80(1):95–101, with permission

Table 4.3 Chest Wall Organ Injury Scale

Grade	Injury type	Description of injury
I	Contusion	Any size
	Laceration	Skin and subcutaneous
	Fracture	*Ribs*: <3 ribs, closed *Clavicle*: Non-displaced, closed
II	Laceration	Skin, subcutaneous, and muscle
	Fracture	*Ribs:* ≥3 adjacent, closed *Clavicle*: Open or displaced *Sternum*: Closed *Scapula*: Body, open, or closed
III	Laceration	Full thickness including pleural penetration
	Fracture	Ribs: Unilateral flail segment (<3 ribs) *Sternum*: Open, displaced, or flail
IV	Laceration	Avulsion of chest wall tissues with underlying rib fractures
	Fracture	*Ribs*: Bilateral flail chest (≥3 on both sides)

Association for the Surgery of Trauma [9]. It is akin to other solid organ grading scales in that it grades injuries from I to V and divides injury patterns into laceration, contusion, and fracture. Furthermore, it is the most comprehensive scoring system in terms of inclusion of additional bones related to the chest wall, such as the sternum, clavicle, and scapula. Additional variables include the number of rib fractures (dichotomized at 3), presence of flail chest, and fracture bilaterally. Finally, the presence and degree of soft tissue injury are included in the scoring system.

Despite its frequent referencing, the Chest Wall Organ Injury Scale has not been validated extensively. When we applied the score to a sample of 385 rib fracture patients from our trauma center, the ROC AUC for predicting respiratory failure, pneumonia, and tracheostomy were 0.61, 0.60, and 0.66, respectively, indicating poor predictive ability [10].

Although the Chest Wall Organ Injury Scale provides an overall assessment of chest wall injury, including adjacent bones and soft tissue, it lacks detailed information regarding specific rib fracture patterns and their relationship to pulmonary physiology. Finally, individual grades are comprised of multiple possible injury patterns, complicating comparison of groups. For example, a grade II injury may involve a displaced clavicle fracture, three or more adjacent rib fractures, an open scapular fracture, or any combination of the three. For this reason, grouping patients using this scale is of limited use when investigating interventions specific to rib fractures.

Rib Fracture Score

The Rib Fracture Score is the most straightforward of the scoring systems, involving only three variables: the number of fractures, the patient age, and the presence of bilateral fractures [11]. It is calculated as (breaks × sides) + age factor (50–60 years = 1 point, 61–70 years = 2 points, 71–80 years = 3 points, and >80 years = 4 points). Accordingly, the score was intended specifically to triage older patients with rib fractures. An additional advantage is that the score can be abstracted readily from national large-scale databases, such as the National Trauma Data Bank. However, subsequent validation of this score revealed poor explanatory ability [10, 12]. This is likely due to the absence of both detailed fracture pattern information and patient demographic and physiologic variables.

Chest Trauma Score

The Chest Trauma Score incorporates 4 parameters: patient age (1–3 points), pulmonary contusion (0–3 points), number of rib fractures (1–3 points), and bilateral fractures (2 points). It ranges from 0 to 11. It was developed using a single-institution sample of 649 patients by Pressley et al. [13] and later validated in another single-center patient population of 1361 patients by Chen et al. [14]. This score is remarkable in that statistical modeling was employed to weight variables differently. For example, bilateral rib fractures are assigned two points instead of one.

In its initial description, significant differences in the outcomes of mortality, ICU admission, mechanical ventilation, and length of stay were observed at thresholds ranging from 4 to 7. The ROC AUCs were not reported. In a larger validation study composed of 1361, differences in outcomes were also observed, including pneumonia and tracheostomy. Again, the ROC AUCs were not reported. When we validated the CTS, AUC ROS when applied to our sample was in the 0.65 range and not substantially different than either the Chest Wall Organ Injury Scale or the Rib Fracture Score.

RibScore

We developed the RibScore with the specific intention of creating a totally radiographic score [10]. Our reasoning is related to a large number of patients who transferred to our institution, for whom only CT information was available, as opposed to often subjective and incomplete clinical information over the phone. We were particularly interested in predicting the need for surgical intervention for rib fractures, and thus transfer to our center, which possesses local expertise in this operation.

We selected six radiographic variables based on both literature and anecdotal evidence of disease severity: ≥6 total fractures; ≥3 fractures with bicortical displacement; radiographic flail segment; at least one fracture in each anatomic region, anterior, lateral, and posterior, defined by anterior and posterior axillary lines; first rib fracture; and bilateral rib fractures. The presence of each parameter was assigned 1 point, such that the score ranged from 0 to 6.

Among our single-institution sample of 385 patients with rib fractures, each of these six parameters was associated significantly with the outcomes of respiratory failure, pneumonia, and tracheostomy. The median RibScore was 1 and ranged from 0 to 5. The ROC AUCs for the three aforementioned outcomes were 0.69, 0.71, and 0.75, indicating good predictive ability. These values improved further to the 0.75–0.85 range when limiting the sample to patients with isolated rib fractures.

Calculation of the RibScore has now become standard operating procedure at our institution; it is reported daily on intensive care unit rounds as readily as the patient's vital signs. The main advantage is that all of the information necessary to calculate it is available from the admission CT chest. Furthermore, of the current rib fracture scoring systems, the RibScore includes the most detail regarding fracture pattern, including anatomic location and degree of displacement. However, it does not include physiologic parameters; as such, it fails to incorporate different phenotypes of the radiographic findings. Furthermore, equal weighting was assigned to each of the variables. In reality, it is unlikely that the presence of any one parameter confers the same risk of pneumonia as the others.

Clinical Rib Score

Most recently, Manay et al. developed a totally clinical score that included eight parameters obtained prospectively on physical exam in 139 patients [15]. This clinical score exists on the opposite end of the spectrum as the RibScore, which includes only radiographic information.

In the Clinical Rib Score, each variable was weighted based on the degree of association with mortality as measured by likelihood ratio (1–2 was assigned 1 point, 2–3 was assigned 2 points, and >3 was assigned 3 points). The variables identified were, in the order of decreasing likelihood ratio, paradoxical respiration, increased depth of respiration, dullness to percussion, rib crunching, tachypnea, chest tenderness, and subcutaneous emphysema.

Although this score will need to be validated among larger samples of rib fracture patients, it represents an important departure from its predecessors in many ways. It was developed prospectively, assigned weighted values to variables based on statistical modeling, and, finally, includes clinical variables. This last distinction is particularly encouraging as few if any rib scoring systems to date share this quality.

Future Directions

Two main tasks are required to refine the topic of chest wall injury scoring. The first is to create and validate a universal nomenclature for chest wall injury. Indeed, a scoring system is only as valid as the parameters used to calculate it. Several national and international organizations are currently in the process of creating such a standardized nomenclature. For example, the Chest Wall Injury Society, using the Delphi method [16], has arrived at expert consensus regarding several previously subjective definitions, such as flail chest, lateral fractures, and displacement. Although no definition will be perfect, the goal will be to assimilate the nomenclature system into everyday clinical and research practice, such that it can be used to develop injury severity scoring systems. The Chest Wall Injury Society nomenclature system is anticipated to be available in the summer of 2018.

Once a standardized nomenclature is agreed upon, the next task will be using it to create a universal scoring system. Aspects from each of the existing, aforementioned systems may be used synthesized into such a score. Preferably, weighted statistical modeling, including strength of association with adverse outcomes, will be used.

In practice, it is likely that two general types of scoring systems will be required to adequately describe chest wall injuries. The first scoring system would involve predominantly anatomic considerations, for example, the current Chest Wall Injury Score and RibScore. This anatomic score would be most useful for tracking the overall epidemiology of chest wall injuries. A second system would involve some combination of both demographic and physiologic parameters, emphasizing the phenotypic expression of the pathology as it relates to the development of complications. This clinical score would likely be most useful when determining which patients are most likely to benefit from specific therapies (e.g., locoregional anesthesia).

One important task remains using a scoring system to delineate who will be best served by operative management of rib fractures. Currently, although most data [17] and expert opinion [18, 19] suggest that the main benefit of surgical stabilization of rib fractures is for patients with flail chest, the surgery is frequently performed for other indications [20]. Further defining subgroups of patients who will derive benefit from this operation will be greatly facilitated by a scoring system that incorporates detailed, non-flail fracture patterns.

Finally, efforts should be made to incorporate additional outcomes beyond mortality and acute respiratory failure. For example, Bugaev et al. found that the magnitude of rib fracture displacement predicted opioid requirements [21]. As mortality from rib fractures continues to decrease, a shift toward quality of life outcomes should occur.

Conclusion

Chest wall injuries should be viewed as a unique group of pathologies that require a dedicated scoring system. Scoring systems are useful for standardizing communication, selecting therapies, and conducting research. In the case of rib fractures, a culture change among providers is necessary, from the general, nebulous statement that "this patient has rib fractures" to a structured, detailed description. Other examples of successfully implemented scoring systems include blunt solid organ injury and cerebrovascular injuries, and chest wall injuries should parallel these developments. Although current chest wall injury scoring systems are limited by a lack of both detail and validation, they should be used nonetheless, as they still represent a structured method for presenting information. Regardless of which system is used, standardized reporting of scores should become ingrained into communication between providers, for example, during daily rounds and patient transfers. Moving forward, the development of both a standard nomenclature and scoring system is urgently needed, in particular in the setting of rapidly emerging surgical therapies, as well as a shift from acute mortality to long-term morbidity from these injuries.

References

1. Pieracci FM, Coleman J, Ali-Osman F, et al. A multicenter evaluation of the optimal timing of surgical stabilization of rib fractures. J Trauma Acute Care Surg. 2018 Jan;84(1):1–10.
2. Tyburski JG, et al. Pulmonary contusions: quantifying the lesions on chest X-ray films and the factors affecting prognosis. J Trauma. 1999;46(5):833–8.
3. Bulger EM, et al. Rib fractures in the elderly. J Trauma. 2000;48(6):1040–6. discussion 1046-7
4. Flagel BT, et al. Half-a-dozen ribs: the breakpoint for mortality. Surgery. 2005;138(4):717–23. discussion 723-5
5. Chien CY, et al. The number of displaced rib fractures is more predictive for complications in chest trauma patients. Scand J Trauma Resusc Emerg Med. 2017;25(1):19.
6. Butts CA, et al. Do simple beside lung function tests predict morbidity after rib fractures? Am J Surg. 2017;213(3):473–7.
7. Carver TW, et al. Vital capacity helps predict pulmonary complications after rib fractures. J Trauma Acute Care Surg. 2015;79(3):413–6.
8. Holcomb JB, et al. Morbidity from rib fractures increases after age 45. J Am Coll Surg. 2003;196(4):549–55.
9. Moore EE, et al. Organ injury scaling. III: chest wall, abdominal vascular, ureter, bladder, and urethra. J Trauma. 1992;33(3):337–9.
10. Chapman BC, et al. RibScore: a novel radiographic score based on fracture pattern that predicts pneumonia, respiratory failure, and tracheostomy. J Trauma Acute Care Surg. 2016;80(1):95–101.
11. Easter A. Management of patients with multiple rib fractures. Am J Crit Care. 2001;10(5):320–7. quiz 328-9
12. Maxwell CA, Mion LC, Dietrich MS. Hospitalized injured older adults: clinical utility of a rib fracture scoring system. J Trauma Nurs. 2012;19(3):168–74. quiz 175-6
13. Pressley CM, et al. Predicting outcome of patients with chest wall injury. Am J Surg. 2012;204(6):910–3. discussion 913-4
14. Chen J, et al. A chest trauma scoring system to predict outcomes. Surgery. 2014;156(4):988–93.
15. Manay P, et al. Studying morbidity and predicting mortality in patients with blunt chest trauma using a novel clinical score. J Emerg Trauma Shock. 2017;10(3):128–33.
16. Dalkey NC, Helmer O. An experimental application of the Delphi method to the use of experts. Manag Sci. 1963;9:458–67.
17. Slobogean GP, et al. Surgical fixation vs nonoperative management of flail chest: a meta-analysis. J Am Coll Surg. 2013;216(2):302–11 e1.

18. Pieracci FM, et al. Consensus statement: surgical stabilization of rib fractures rib fracture colloquium clinical practice guidelines. Injury. 2017;48(2):307–21.
19. Kasotakis G, et al. Operative fixation of rib fractures after blunt trauma: a practice management guideline from the eastern Association for the Surgery of trauma. J Trauma Acute Care Surg. 2017;82(3):618–26.
20. Kane E, et al. Quantifying and exploring the recent national increase in surgical stabilization of rib fractures. J Trauma Acute Care Surg. 2017 Dec;83(6):1047–52.
21. Bugaev N, et al. Magnitude of rib fracture displacement predicts opioid requirements. J Trauma Acute Care Surg. 2016;81(4):699–704.

Medical Management of Rib Fractures

5

Tashinga Musonza and S. Rob Todd

Rib fractures are reported in greater than 10% of all trauma admissions [1]. A staggering 60% of patients admitted with thoracic trauma have rib fractures. Bulger et al. documented that patients older than 65 years of age with two or more rib fractures were more likely to die as compared to similarly injured patients between the ages of 15 and 64. The risk of mortality with each additional rib fracture increases by 19% in this population. Similarly, the risk of nosocomial pneumonia has been documented to be as high as 29% [2]. More recent studies have shown an increase in mechanical ventilator days and hospital length of stay in patients older than 45 years of age with four or more rib fractures. Mortality is likewise correlated with an increasing number of rib fractures [3, 4].

Despite the overwhelming morbidity and mortality associated with rib fractures, their medical management remains relatively simplistic. Likewise, rib fracture clinical care pathways or guidelines are yet to be universally adopted in spite of data showing improved outcomes with such utilization.

Initial Evaluation

Though rib fractures may uncommonly be seen in forceful coughing, this chapter will focus on trauma patients being admitted to the hospital [5]. The initial evaluation and management hinge on the diagnostic evaluation and resuscitation as dictated by the Advanced Trauma Life Support (ATLS) protocols. Prioritization is geared toward airway, breathing, and circulation [6, 7]. The overall trauma burden and patient physiology subsequently guide patient management and disposition. The mechanism of trauma, forces involved, and associated complications should be taken into account.

T. Musonza · S. R. Todd (✉)
Michael E. DeBakey Department of Surgery, Baylor College of Medicine, Houston, TX, USA
e-mail: tashinga.musonza@bcm.edu; srtodd@bcm.edu

M. de Moya, J. Mayberry (eds.), *Rib Fracture Management*,
https://doi.org/10.1007/978-3-319-91644-6_5

A focused and then complete physical examination is warranted upon initial contact with the patient. The presence of chest wall pain, hematoma or ecchymosis, flail chest, or significant chest or neck crepitus hints toward clinically significant thoracic trauma and, as such, the possibility of multiple rib fractures. Only a flail chest is specific for rib fractures, while other findings are highly suggestive.

Chest radiography is an adjunct to the primary survey of trauma patients. Computed tomography (CT) is reported to be superior to chest radiography in detecting displaced or posterior rib fractures [8]. Computed tomography also invariably detects all if not most of the rib fractures in addition to associated injuries [9]. That being said, chest radiography provides an expedient method for ruling out life-threatening thoracic injuries including open pneumothorax, massive hemothorax, and flail chest and potentially life-threatening injuries such as thoracic aortic disruption, tracheobronchial disruption, traumatic diaphragmatic tear, and pulmonary contusion.

As part of this initial evaluation, the presence of a first or second rib fracture or three or more rib fractures should raise the index of suspicion for blunt cardiac injury. This high-energy fracture pattern mandates blunt cardiac injury screening to include a serum troponin level and an electrocardiogram [10].

It is worth mentioning that apart from the thorax, associated injuries may be intra-abdominal or intra-calvarium or involve the upper extremity [11]. Management of these other injuries will not be discussed here.

Management

Patient Disposition

Single rib fractures without associated injuries may be safely managed on an outpatient basis in select patients. The adequacy of pain control, reliable follow-up, frailty index, and associated morbidity should be considered. Patients with a higher frailty index, significant morbidity, or poorly controlled pain may require hospitalization or admission to a rehabilitation facility [12].

Once the diagnosis of multiple rib fractures has been made, the patient's age, trauma burden, and number of ribs fractured are integral to the patient's disposition. Patient's older than 45 with four or more rib fractures should be admitted to a monitored setting [3]. This may either be an intensive care unit (ICU) or an intermediate care unit where the patient's respiratory status may be closely monitored. Patients older than 65 with two or more rib fractures should be treated in a similar manner.

Apart from findings such as tachypnea, hypoventilation, or hypoxia, bedside spirometry provides an excellent objective assessment of the patient's respiratory mechanics. Carver et al. reported that patients with fractured ribs and a vital capacity of less than 30% had increased pulmonary complications. A 10% increase in the vital capacity was associated with a 36% decrease in the likelihood of pulmonary complications [13].

Figure 5.1 depicts our institution's rib fracture guideline. Patient age and the number of ribs fractured play crucial roles in patient disposition. Additionally, an incentive spirometry goal of 15 mL/kg ideal body weight is utilized to further guide disposition.

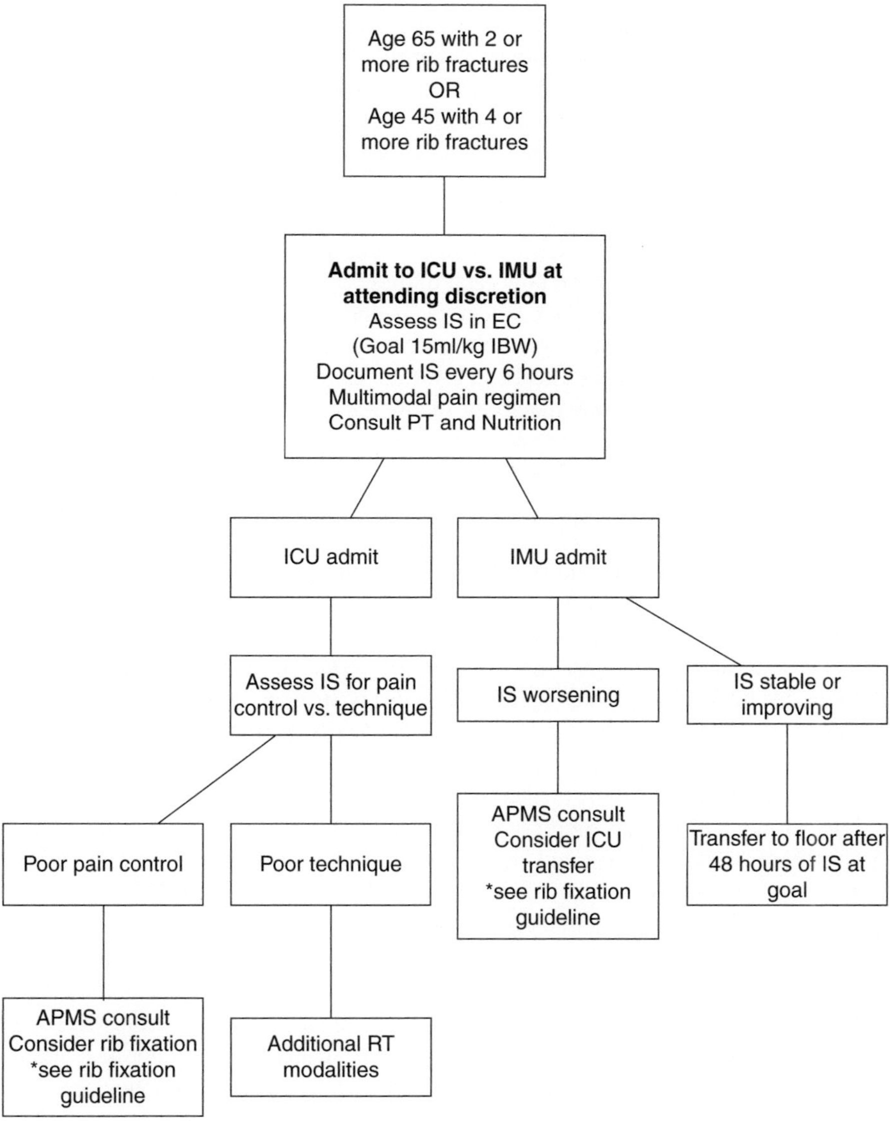

Fig. 5.1 Ben Taub Hospital rib fracture guideline. *ICU* intensive care unit, *IMU* intermediate care unit, *APMS* acute pain management services, *EC* emergency center, *PT* physical therapy. *Rib fixation is addressed in another chapter

Pulmonary Care

The primary goal of pulmonary care in patients with rib fractures is to avoid respiratory failure and the need for subsequent intubation and mechanical ventilation. Secondary end points include effective pulmonary clearance, decreasing atelectasis, prevention of pneumonia, decreased hospital length of stay, and decreased mechanical ventilator days for already intubated patients. Mechanical ventilator-induced diaphragmatic dysfunction secondary to atrophy has been reported. The oxidative stress and mitochondrial changes seen in this condition may be due to the direct effects of positive pressure ventilation, underlying inflammatory process, negative protein balance, or respiratory muscle inactivity [14–18].

Many institutional rib fracture guidelines (including ours) focus heavily on volume expansion techniques. These include coughing, deep breathing exercises, incentive spirometry, and intermittent and continuous positive airway pressure. There is a lack of consensus on how each of these interventions should be administered. In systematic reviews, the effect of incentive spirometry on postoperative pulmonary complications after cardiopulmonary bypass and upper abdominal surgery did not show clear benefits [19]. However, the theoretical benefit of avoiding atelectasis and diaphragmatic inactivity seems logical and makes this intervention attractive. It is also low cost. Thus, incentive spirometry should be mandated in all patients with rib fractures. It should be administered around the clock during waking hours. All this being said, there is lack of data stipulating how to explicitly prescribe incentive spirometry.

In addition to volume expansion, we recommend aggressive early mobilization, sitting up, and getting out of the bed. The increased activity promotes deep breathing thus offsetting atelectasis. In the supine position, the functional residual capacity is reduced by at least 0.8 L due to the cephalad displacement of the diaphragm by abdominal contents [20]. This leads to decreased lung compliance which may already be an issue due to underlying rib fracture complications such as pulmonary contusion, pneumothorax, and hemothorax. The resulting ventilation-perfusion mismatch is further exacerbated during mechanical ventilation. While supine, perfusion is increased to the dependent lung regions though ventilation remains evenly distributed. Early mobilization may play a crucial role in mitigating these factors.

A review of patients with flail chest from the National Trauma Data Bank (NTDB) reported ventilatory support in 59% of this population [21]. In this setting, decreasing ventilator days and preventing ventilator-associated pneumonia (VAP) are imperative. Daily sedation holidays and spontaneous breathing trials (SBTs) as part of a VAP bundle are crucial in preventing VAP. Daily SBTs cannot be overemphasized. The use of daily SBTs is associated with decreased ventilator and ICU days, as well as a decreased 1-year mortality with a 0.68 hazard ratio [22]. Additionally, our VAP bundle consists of elevating the head of bed to 30 degrees, twice a day oral rinsing with 0.12% chlorohexidine, and frequent oral or endotracheal suctioning. Early liberation from the ventilator is however the most effective way to prevent VAP.

The avoidance of barotrauma, control of underlying inflammatory process, and mitigating malnutrition probably impact the duration of ventilator-free days [17, 18, 23, 24]. We will discuss nutrition management in a subsequent section.

Fluid Management

Trauma patients often require aggressive intravascular volume resuscitation. Fluid therapy should be judicious but adequate to maintain an optimum intravascular volume. Clinicians should strive for a net even or net negative fluid balance over the course of the patient's hospital stay. This is especially true in patients with pulmonary contusions—the presence of lung parenchymal damage with edema and hemorrhage in the absence of a pulmonary laceration. In such instances, a net positive fluid balance may worsen pulmonary edema, hence ventilation-perfusion mismatch thus contributing to hypoxic vasoconstriction [25]. This may worsen cardiac dysfunction due to elevated pulmonary artery pressures, thereby negatively impacting the ability to wean at-risk patients from mechanical ventilation.

The need to administer intravascular volume is best guided by dynamic parameters such as stroke volume variability (SVV), pulse pressure variability (PPV), or inferior vena cava (IVC) ultrasound measurements in patients on the mechanical ventilator. We recommend the routine use of SVV, PPV, or IVC measurements in assessing fluid volume responsiveness in such patients [26, 27]. A passive leg raise is also useful in assessing fluid responsiveness; however, it is difficult to administer in trauma patients due to the presence of concomitant abdominal and/or extremity injuries.

Daily fluid requirements should be calculated based on the patient's dry weight. The type of fluid therapy is dictated by the indication. For resuscitative volume, we advocate crystalloids instead of albumin. This is in keeping with the SAFE trial which showed no mortality difference when 4% albumin and normal saline were individually utilized as volume expanders in the ICU [28]. Crystalloids should be used for maintenance therapy when enteral intake is not permissible. Additives such as electrolytes may be utilized as influenced by the patient's physiologic state.

Pain Management

As depicted in Fig. 5.2, pain is a driving force behind rib fracture complications. Thus, pain control is a cornerstone in the management of rib fractures [29]. We will discuss different pain control modalities and their relative application.

Multimodal Pain Regimen

Ketorolac, a nonsteroidal anti-inflammatory drug (NSAID), has been shown to decrease both ventilator days and the rates of pneumonia in ICU patients with rib fractures [30]. Nonsteroidal anti-inflammatory drugs may be given around the clock in addition to scheduled acetaminophen. Intravenous formulations are also available

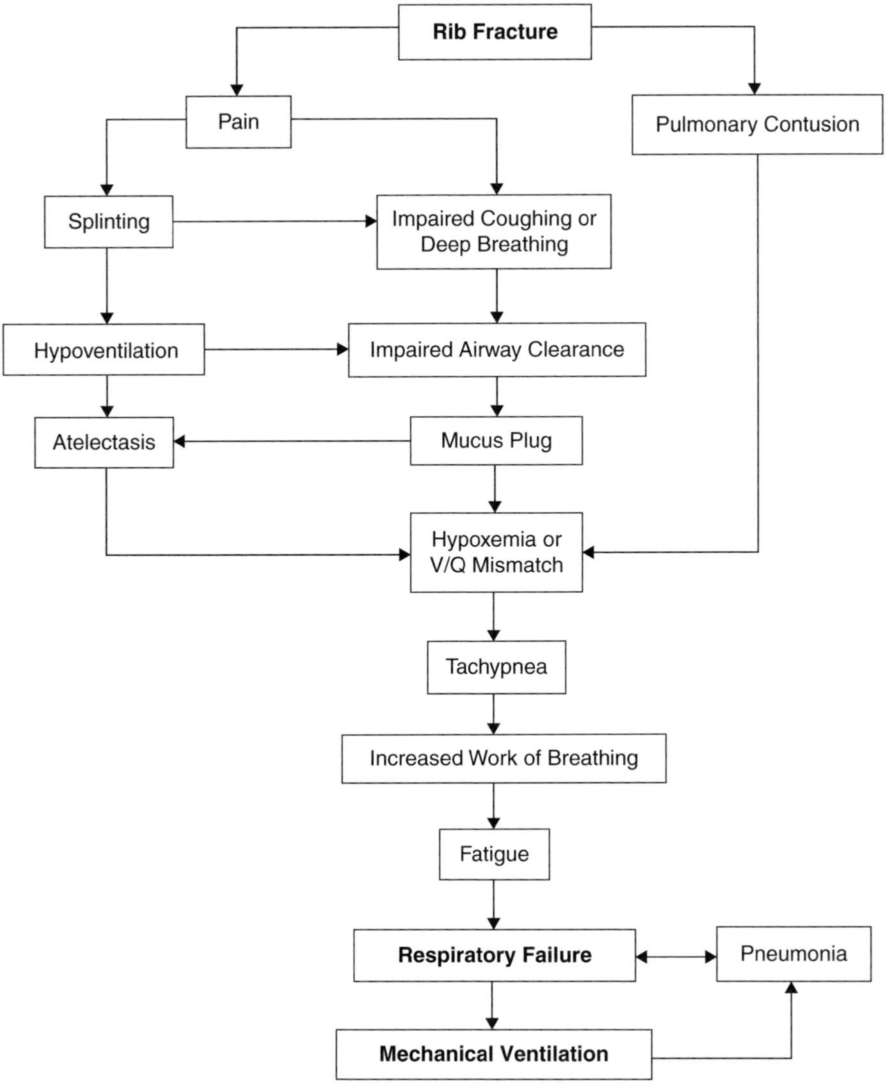

Fig. 5.2 The vicious cycle of rib fractures

for patients unable to tolerate the enteral route. Renal function should be evaluated prior to scheduling NSAID therapy. Given the high-energy impact required to cause rib fractures, it is unlikely that the associated pain will be controlled by NSAIDs and acetaminophen alone. Our multimodal regimen consists of scheduled 1000 mg acetaminophen every 6 h, 300 mg gabapentin twice a day, 100 mg tramadol every 6 h, and NSAID therapy (Fig. 5.3). Either naproxen or ketorolac can be used if the (estimated glomerular filtration rate) eGFR is greater than 50. Gabapentin has been reported to produce analgesia and opioid-sparing effects in the postoperative period [31, 32]. We use opioid-based medication for rescue purposes only.

HARRISHEALTH SYSTEM TRAUMA/CRITICAL CARE SURGERY **DEPARTMENTAL GUIDELINES AND PROCEDURES**	Procedure NO: Page Number: 1 of 2 Effective Date: 03/17

TITLE: MANAGEMENT OF ACUTE PAIN FOR ACUTE CARE AND TRAUMA SURGERY PATIENTS

PURPOSE: This document is intended to provide guidance to physicians and mid-level providers as to standard of care management will in the treatment of acute pain following surgery or injury.

GUIDELINES/PROCEDURES STATEMENT:

Scope: These recommendations were developed in collaboration with the Anesthesia Department Acute Pain Service and Acute Care Surgical Service.

1. All patients should receive multimodal SCHEDULED analgesia therapy in patients with n*ormal renal and hepatic function.* This consists of scheduled acetaminophen, anti-inflammatory therapy, gabapentin and tramadol.
   - **a. Acetaminophen Therapy**
     - a. Acetaminophen 1gm PO or IV q6hr
       - → use IV ONLY if NPO/Not Tolerating PO × 24 hours
   - **b. NSAID Therapy** (*ensure eGFR > 50*)
     - a. Naproxen 500mg PO q12hr *OR* Ibuprofen 800mg PO q8hr (max 3200 mg/day)
     - b. Ketorolac (Toradol®) 15mg IV q 6hrs scheduled x 96hrs maximum
   - **c. Gabapentin**
     - a. Gabapentin 300mg po Daily x 1 day then 300mg PO BID (*inpatient use only*)
   - **d. Tramadol 100 mg PO q6hr**
     - → In patients < 70kg reduce dose to 50mg PO q6hr
2. Patients should receive opioid-based pain medication for rescue purposes. Opioid tolerant patients may reqire schedduled narcotic to prevent withdrawal.
   - a. **PO Rescue Medications (if tolerating PO or enteral nutrition)**
     - 1. Oxycodone (Roxicodone®): 5mg q 4hr prn moderate pain
     - 2. Oxycodone (Roxicodone®): 10 mg q4h prn severe pain
   - b. **IV Rescue Medications:**
     - 1. Morphine 2mg IV q 3hrs PRN severe pain
     - 2. Dilaudid 0.5mg IV q 3hrs PRN severe pain
3. **Baseline PCA Settings:**
   - a. Morphine PCA: 1mg q10min, no basal rate
   - b. Dilaudid PCA: 0.2mg q10min, no basal rate
4. **Bowel Regimen:** All patients receiving opioid therapy should also receive a bowel regimen.
   - a. Bisacodyl 10 mg po daily (*please refer to Bowel Regimen Trauma Guidelines*)
5. **Anti-Emetic Therapy**
   - a. Promethazine (Phenergan ®) 6.25-25 mg IV, PO or PR q6hr PRN N/V

     OR

     Ondansetron (Zofran®) 4mg IV or PO q6hr scheduled if N/V exists

Fig. 5.3 Ben Taub Hospital multimodal pain regimen

Intravenous opioids may be required to control rib fracture pain. They have a rapid onset of action and predictable duration and therefore are easy to dose. However, they should be used with caution due to the likelihood of respiratory depression and other opioid-associated complications such dysphoria, sedation, constipation, urinary retention, and nausea. If one is to utilize intravenous opioids, patient-controlled analgesia (PCA) is recommended as it provides analgesia in a timely manner and is less likely to cause untoward sedation (with appropriate dosing). A daily bowel regimen should be prescribed concurrently to mitigate the constipative effects of opioids.

Opioid-sparing adjuncts such as ketamine and lidocaine drips may be considered during the management of rib fractures. Wang et al. compared low-dose ketamine infusion and a morphine PCA to a morphine PCA alone in the postoperative period. The addition of low-dose ketamine resulted in better pain control and decreased opioid use [33]. That being said, questions remain as to the dosage and duration of ketamine infusions. The use of a lidocaine infusion following abdominal surgery has likewise been reported to decrease postoperative pain. However, the same meta-analysis did not find consistent results pertaining to the effect of lidocaine infusion on the return of bowel function, opioid requirements, and postoperative nausea or vomiting [34].

If the pain of rib fractures is poorly controlled with a multimodal pain regimen and opioids, regional anesthesia should strongly be considered. Regional blocks have been reported to be superior to systemic opioids and produce less systemic side effects [29].

Regional Anesthesia

Epidural Anesthesia. The use of epidural anesthesia has been shown to decrease ventilator days in patients with multiple rib fractures [35]. Belger et al. documented similar results in addition to a decrease in the rate of nosocomial pneumonia when epidural anesthesia was used instead of intravenous opioids [36]. The use of thoracic epidural anesthesia may be contraindicated in some patients due to coagulopathy, spine fractures, or hypotension, which may be common in trauma patients. Combative or confused patients may not be able to lie still for the procedure.

With epidural anesthesia, the clinician should monitor the patient for hypotension and less commonly spinal hematoma or epidural abscess. Motor and sensory function of the lower extremities should likewise be routinely assessed. Epidural anesthesia has the advantage of avoiding opioids thus eliminating the risk of associated complications. Hypotension is still a problem even with ropivacaine or bupivacaine; however, the concentration or rate of epidural anesthetic infusion can be decreased. Epidural anesthesia should be managed by appropriately trained personnel, preferably a pain management specialist.

Paravertebral Nerve Block. The pain resulting from unilateral rib fractures may be well controlled with paravertebral nerve blocks. The risk of hypotension or respiratory depression seen with intravenous opioids or epidural anesthesia is significantly decreased. Pain control with paravertebral blocks has been documented to be equivalent if not superior to epidural anesthesia [37]. Additionally, paravertebral catheters may be left in place even as an outpatient.

Intercostal Nerve Block. Intercostal nerve blocks are appropriate for unilateral and/or a limited number of rib fractures. Their use is limited by the duration of action of the anesthetic used thus often necessitating the need for further blockade if pain relief is to be maintained. This may limit the practicality of intercostal nerve blocks.

Intrapleural Infiltration. The data on the infusion of an anesthetics into the pleural space has not had convincing results and is thus not recommended.

Venous Thromboembolism Prophylaxis

Patients with rib fractures often have concomitant injuries posing a limit to their mobility and functional status. They often have the triad of coagulopathy, stasis, and endothelial injury due to trauma; therefore, they are at increased risk of developing deep vein thrombosis (DVT).

Early mobilization and consultation with physical therapy are of utmost importance.

Apart from mobilization, mechanical and chemical methods of thromboprophylaxis should be employed. Intermittent pneumatic compression devices are the recommended mechanical means. In the CLOTS 1 trial, graduated compression stockings were not beneficial in preventing venous thromboembolic events (VTE). Rather, there was an increase in adverse events such as skin necrosis, breakdown, and ulceration [38]. Intermittent pneumatic compression devices should be utilized unless the underlying trauma precludes them. Randomized trials have proven that pharmacologic prophylaxis with either low molecular weight heparin, unfractionated heparin, or fondaparinux is superior to mechanical methods alone in preventing VTE [39]. Thus, there is insufficient evidence to recommend mechanical thromboprophylaxis alone; therefore, we prescribe enoxaparin 30 mg subcutaneously every 12 h. If the creatinine clearance is less than 30 mL/min, we prescribe heparin 5000 units subcutaneously every 8 h.

Nutritional Support

Trauma patients with multiple rib fractures are often critically ill, and as such, the underlying acute physiologic stress response may lead to anorexia and/or delayed gastric emptying. Oral intake may be hindered due to a variety of reasons such as facial trauma, altered mental status, and mechanical ventilation. The resulting catabolic state may worsen outcomes as well as propagate diaphragmatic dysfunction seen in ventilator-dependent patients [23, 40]. Estimating accurate energy requirements in acute illness is often difficult. A dietician should be involved in the care of all critically ill patients. We advocate early enteral nutrition within 48 h of hospital admission. A study by Artinian et al. documented decreased ICU and hospital mortality with the initiation of nutritional support within 2 days of admission [41].

If oral intake is not permissible, we recommend the placement of a post-ligament of Treitz feeding tube in patients who are at high risk for aspiration (or in the stomach for those who are not) [42, 43]. We do not check gastric residual volumes in keeping with the REGANE and NUTRIREA 1 trials [44, 45]. Holding tube feeding for gastric residuals greater than 200 mL has not been shown to decrease the rates of aspiration pneumonia. In fact, studies done by McClave et al. reported no difference in the rates of aspiration when tube feedings were held for residuals greater than 200 mL versus 400 mL [46]. Checking gastric residuals results in frequent feeding interruptions and thus inadequate caloric intake. Tube feeds should be adjusted based on clinical findings such bloating, intractable nausea, abdominal distension, and pain. Pro-motility agents such as metoclopramide or erythromycin should be considered for those patients showing clinical signs of delayed gastric emptying. Total parenteral nutrition should be considered on hospital day 7 if full enteral nutrition is not achieved (or anticipated to be achieved by hospital days 10–14) [47].

Complications

A key component in rib fracture management is the prevention and management of rib fracture complications. Some of these deserve special attention. As previously mentioned, life-threatening complications such as pneumothorax and hemothorax are encountered in the immediate aftermath of severe chest trauma [48]. These often require a tube thoracostomy. Pain is probably the most debilitating complication (which has already been discussed). Patients with rib fractures often have pulmonary contusions which may manifest with tachypnea, hypoxemia, and/or increased work of breathing. The presence of pulmonary contusions impacts fluid volume therapy (previously discussed). Worsening respiratory status in a patient with known pulmonary contusion warrants further diagnostic imaging. The risk of pneumonia as well as acute respiratory distress syndrome (ARDS) may be higher in this population [49]. Respiratory failure and pneumonia result from the interaction of a host of factors as depicted in Fig. 5.2.

Conclusion

The significant impact of rib fractures on morbidity and mortality is unquestionable. As such, the management strategies are multimodal to include pain control, pulmonary care, early mobilization, nutritional therapy, appropriate fluid management, and DVT prophylaxis. The use of clinical care pathways or guidelines has resulted in improved outcomes. A clinical care pathway utilized for patients above the age of 45 years with four or more rib fractures resulted in decreased infectious morbidity, mortality, length of stay, and ventilator days. Figure 5.1 depicts the rib fracture guideline algorithm currently used at our institution. At the core of this guideline is identifying high-risk patients and appropriately triaging them to an ICU or monitored care unit. From there, the tactic is multimodal as aforementioned.

References

1. Bergeron E, et al. Elderly trauma patients with rib fractures are at greater risk of death and pneumonia. J Trauma. 2003;54(3):478–85.
2. Bulger EM, et al. Rib fractures in the elderly. J Trauma. 2000;48(6):1040–6. discussion 1046-7
3. Todd SR, et al. A multidisciplinary clinical pathway decreases rib fracture-associated infectious morbidity and mortality in high-risk trauma patients. Am J Surg. 2006;192(6):806–11.
4. Holcomb JB, et al. Morbidity from rib fractures increases after age 45. J Am Coll Surg. 2003;196(4):549–55.
5. Sano A, Tashiro K, Fukuda T. Cough-induced rib fractures. Asian Cardiovasc Thorac Ann. 2015;23(8):958–60.
6. Newgard CD, et al. Emergency medical services intervals and survival in trauma: assessment of the "golden hour" in a north American prospective cohort. Ann Emerg Med. 2010;55(3):235–246.e4.
7. Teixeira PG, et al. Preventable or potentially preventable mortality at a mature trauma center. J Trauma. 2007;63(6):1338–46. discussion 1346-7
8. Pulley BR, et al. Utility of three-dimensional computed tomography for the surgical management of rib fractures. J Trauma Acute Care Surg. 2015;78(3):530–4.
9. Murphy CEt, et al. Rib fracture diagnosis in the Panscan era. Ann Emerg Med. 2017 Dec;70(6):904–9.
10. Ismailov RM, et al. Trauma associated with cardiac dysrhythmias: results from a large matched case-control study. J Trauma. 2007;62(5):1186–91.
11. Al-Hassani A, et al. Rib fracture patterns predict thoracic chest wall and abdominal solid organ injury. Am Surg. 2010;76(8):888–91.
12. Ho SW, et al. Risk of pneumonia in patients with isolated minor rib fractures: a nationwide cohort study. BMJ Open. 2017;7(1):e013029.
13. Carver TW, et al. Vital capacity helps predict pulmonary complications after rib fractures. J Trauma Acute Care Surg. 2015;79(3):413–6.
14. Larsson L, Friedrich O. Critical illness myopathy (CIM) and ventilator-induced diaphragm muscle dysfunction (VIDD): acquired myopathies affecting contractile proteins. Compr Physiol. 2016;7(1):105–12.
15. Mankowski RT, et al. Intraoperative hemidiaphragm electrical stimulation reduces oxidative stress and upregulates autophagy in surgery patients undergoing mechanical ventilation: exploratory study. J Transl Med. 2016;14(1):305.
16. Vassilakopoulos T, Petrof BJ. A stimulating approach to ventilator-induced diaphragmatic dysfunction. Am J Respir Crit Care Med. 2017;195(3):281–2.
17. Yang Y, et al. Endotoxemia accelerates diaphragm dysfunction in ventilated rabbits. J Surg Res. 2016;206(2):507–16.
18. Zhu X, et al. The role of calpains in ventilator-induced diaphragm atrophy. Intensive Care Med Exp. 2017;5(1):14.
19. Hall JC, et al. Prevention of respiratory complications after abdominal surgery: a randomised clinical trial. BMJ. 1996;312(7024):148–52. discussion 152-3
20. Wahba RW. Perioperative functional residual capacity. Can J Anaesth. 1991;38(3):384–400.
21. Gage A, et al. The effect of epidural placement in patients after blunt thoracic trauma. J Trauma Acute Care Surg. 2014;76(1):39–45. discussion 45-6
22. Girard TD, et al. Efficacy and safety of a paired sedation and ventilator weaning protocol for mechanically ventilated patients in intensive care (awakening and breathing controlled trial): a randomised controlled trial. Lancet. 2008;371(9607):126–34.
23. Jaber S, et al. Clinical review: ventilator-induced diaphragmatic dysfunction--human studies confirm animal model findings! Crit Care. 2011;15(2):206.
24. Petrof BJ, Jaber S, Matecki S. Ventilator-induced diaphragmatic dysfunction. Curr Opin Crit Care. 2010;16(1):19–25.

25. Pinsky MR. Cardiovascular issues in respiratory care. Chest. 2005;128(5):592S–7S.
26. Huang CC, et al. Prediction of fluid responsiveness in acute respiratory distress syndrome patients ventilated with low tidal volume and high positive end-expiratory pressure. Crit Care Med. 2008;36(10):2810–6.
27. Marik PE, et al. Dynamic changes in arterial waveform derived variables and fluid responsiveness in mechanically ventilated patients: a systematic review of the literature. Crit Care Med. 2009;37(9):2642–7.
28. Finfer S, et al. A comparison of albumin and saline for fluid resuscitation in the intensive care unit. N Engl J Med. 2004;350(22):2247–56.
29. Karmakar MK, Ho AM. Acute pain management of patients with multiple fractured ribs. J Trauma. 2003;54(3):615–25.
30. Yang Y, et al. Use of ketorolac is associated with decreased pneumonia following rib fractures. Am J Surg. 2014;207(4):566–72.
31. Ho KY, Gan TJ, Habib AS. Gabapentin and postoperative pain--a systematic review of randomized controlled trials. Pain. 2006;126(1–3):91–101.
32. Mathiesen O, Moiniche S, Dahl JB. Gabapentin and postoperative pain: a qualitative and quantitative systematic review, with focus on procedure. BMC Anesthesiol. 2007;7:6.
33. Wang L, et al. Ketamine added to morphine or hydromorphone patient-controlled analgesia for acute postoperative pain in adults: a systematic review and meta-analysis of randomized trials. Can J Anaesth. 2016;63(3):311–25.
34. Weibel S, et al. Efficacy and safety of intravenous lidocaine for postoperative analgesia and recovery after surgery: a systematic review with trial sequential analysis. Br J Anaesth. 2016;116(6):770–83.
35. Mohta M, et al. Prospective, randomized comparison of continuous thoracic epidural and thoracic paravertebral infusion in patients with unilateral multiple fractured ribs--a pilot study. J Trauma. 2009;66(4):1096–101.
36. Bulger EM, et al. Epidural analgesia improves outcome after multiple rib fractures. Surgery. 2004;136(2):426–30.
37. Shapiro BS, et al. Comparative analysis of the paravertebral analgesic pump catheter with the epidural catheter in elderly trauma patients with multiple rib fractures. Am Surg. 2017;83(4):399–402.
38. Dennis M, et al. Effectiveness of thigh-length graduated compression stockings to reduce the risk of deep vein thrombosis after stroke (CLOTS trial 1): a multicentre, randomised controlled trial. Lancet. 2009;373(9679):1958–65.
39. Lilly CM, et al. Thrombosis prophylaxis and mortality risk among critically ill adults. Chest. 2014;146(1):51–7.
40. Alberda C, et al. The relationship between nutritional intake and clinical outcomes in critically ill patients: results of an international multicenter observational study. Intensive Care Med. 2009;35(10):1728–37.
41. Artinian V, Krayem H, DiGiovine B. Effects of early enteral feeding on the outcome of critically ill mechanically ventilated medical patients. Chest. 2006;129(4):960–7.
42. Taylor SJ, et al. A randomised controlled feasibility and proof-of-concept trial in delayed gastric emptying when metoclopramide fails: we should revisit nasointestinal feeding versus dual prokinetic treatment: achieving goal nutrition in critical illness and delayed gastric emptying: trial of nasointestinal feeding versus nasogastric feeding plus prokinetics. Clin Nutr ESPEN. 2016;14:1–8.
43. Taylor SJ, Manara AR, Brown J. Treating delayed gastric emptying in critical illness: metoclopramide, erythromycin, and bedside (cortrak) nasointestinal tube placement. JPEN J Parenter Enteral Nutr. 2010;34(3):289–94.
44. Montejo JC, et al. Gastric residual volume during enteral nutrition in ICU patients: the REGANE study. Intensive Care Med. 2010;36(8):1386–93.
45. Casaer MP, Van den Berghe G. Nutrition in the acute phase of critical illness. N Engl J Med. 2014;370(13):1227–36.

46. McClave SA, et al. Poor validity of residual volumes as a marker for risk of aspiration in critically ill patients. Crit Care Med. 2005;33(2):324–30.
47. SA MC. Guidelines for the provision and assessment of nutrition support therapy in the adult Criticall ill patient: Society of Critical Care Medicine (SCCM) and American Society for Parenteral and Enteral Nutrition (a.S.P.E.N). JPEN J Parenter Enteral Nutr. 2009;33(3):277–316.
48. Sirmali M, et al. A comprehensive analysis of traumatic rib fractures: morbidity, mortality and management. Eur J Cardiothorac Surg. 2003;24(1):133–8.
49. Allen GS, Cox CS Jr. Pulmonary contusion in children: diagnosis and management. South Med J. 1998;91(12):1099–106.

Lung Contusion Management: Invasive and Noninvasive

6

Umar Bhatti and Lena M. Napolitano

History

The first description of lung injury not associated with chest wall injury was in 1761 by the Italian anatomist Giovanni Battista Morgagni, with a subsequent report by R.W. Smith of Dublin in 1840 [1]. The term "pulmonary contusion" is thought to have been coined by French military surgeon Guillaume Dupuytren in the nineteenth century [2]. Initial published descriptions of lung/pulmonary contusion in the medical literature began in 1965 [3–5]. Lung contusion was described as follows: "This implies the extravasation of blood into alveoli and bronchi as a result of injury. Diffusion is interfered with, ventilation/perfusion ratios are upset, large right-to-left shunts can occur, and ventilation may be greatly impaired" [6].

Epidemiology

Thoracic trauma occurs in more than 50% of blunt trauma patients [7]. Pulmonary contusion is a common finding after blunt chest trauma. Lung contusion is present in 30–75% of all blunt thoracic trauma patients. The most common mechanisms of injury include passengers or pedestrians in motor vehicle and motorcycle crashes and falls. But pulmonary contusion rarely occurs in isolation. Pulmonary contusion occurs in combination with other thoracic injuries, and therefore the independent disease burden of lung contusion

U. Bhatti · L. M. Napolitano (✉)
Acute Care Surgery, Trauma, Burn, Surgical Critical Care, Department of Surgery, University of Michigan Health System, Ann Arbor, MI, USA
e-mail: bhattiu@umich.edu; lenan@umich.edu

M. de Moya, J. Mayberry (eds.), *Rib Fracture Management*,
https://doi.org/10.1007/978-3-319-91644-6_6

is difficult to determine. Complications such as pneumonia and adult respiratory distress syndrome (ARDS) occur in up to 50% of patients with lung contusion.

Pathophysiology

Blunt traumatic thoracic injuries cause damage to the pulmonary parenchyma and alveoli, with blood and fluid accumulation in the lung tissue. This interferes with pulmonary gas exchange at the alveolar level, resulting in hypoxemia and hypercarbia. The pathophysiology of lung contusion includes ventilation/perfusion mismatch, increased intrapulmonary shunt, increased lung water and pulmonary edema, and significantly decreased pulmonary compliance.

Increased alveolar-capillary permeability and inflammation and hemorrhage in the contused lung is common, and a surrounding area of pulmonary edema in the lung adjacent to the lung contusion site is also common. Fluid, blood, and protein accumulation in the alveoli leads to alveolar collapse with resultant impaired gas exchange. Alveolar consolidation leads to inflammation, increased alveolar-capillary membrane permeability with pulmonary edema, ventilation/perfusion mismatch, increased intrapulmonary shunt, and decreased lung compliance. Furthermore, surfactant depletion can occur with lung contusion which further results in alveolar flooding with subsequent hypoxemia.

Lung contusion has an unpredictable pulmonary clinical course, with some cases resolving with only supportive care and supplemental oxygen and other cases resulting in severe hypoxemia and severe acute respiratory distress syndrome requiring advanced mechanical ventilation strategies. The physiologic consequences of alveolar hemorrhage and pulmonary parenchymal destruction typically manifest themselves within hours of injury and usually resolve within approximately 7 days. Clinical symptoms, including hypoxemia and hypercarbia, peak at about 72 h after injury.

Diagnosis

Diagnosing pulmonary contusion requires a clinical history that details information about the inciting traumatic event, a pertinent physical examination, and clues from radiographic tests. All trauma patients who sustained chest wall trauma should be evaluated for pulmonary contusion. Laboratory findings are of limited use because arterial blood gas levels may not show any abnormality (hypoxemia or hypercapnia) in the early course of pulmonary contusion.

Initial Chest Radiograph

A portable chest radiograph is recommended in the initial evaluation of trauma patients with possible thoracic injury, but it has limited utility in evaluation of the acute injury of lung contusion. The presence of hemothorax or pneumothorax may obscure the findings of lung contusion on chest radiograph. The initial chest radiograph may also be normal, as it often takes 6–48 h for lung contusions to become fully apparent on chest radiograph.

Lung Ultrasonography

The overall sensitivity of 94.6% for ultrasound and 27% for initial CXR to diagnose lung contusion provides sufficient evidence for ultrasonography to diagnose lung contusion in emergency settings [8]. Lung ultrasound can be used to identify lung contusion by examining for *B-lines* (which indicate pulmonary edema) to help "rule out" lung contusion and *C-lines* (confluent consolidations or "hepatization" which identify pulmonary parenchymal consolidation) to "rule in" lung contusion (Fig. 6.1) [9]. Lung ultrasound provides a rapidly available point-of-care technique to evaluate acute lung injury, including pneumothorax, hemothorax, and lung contusions. It is especially valuable since it is easy to learn and less technically demanding than other ultrasound procedures [10].

Thoracic CT Scan Imaging

Thoracic CT scan imaging is highly sensitive in detecting thoracic injuries after blunt chest trauma and is superior to chest radiograph in confirming lung contusions, pneumothorax, and hemothorax [11, 12]. Lung contusion is typically identified by thoracic CT imaging as focal, non-segmental areas of parenchymal opacification, usually peripheral in location (Fig. 6.2). Lung contusion is more commonly identified in the lower lobes and more posteriorly, but this can differ in lateral-impact trauma. Thoracic CT scan imaging also provides excellent examination of rib fractures and flail chest, and bony reformats can be very helpful in

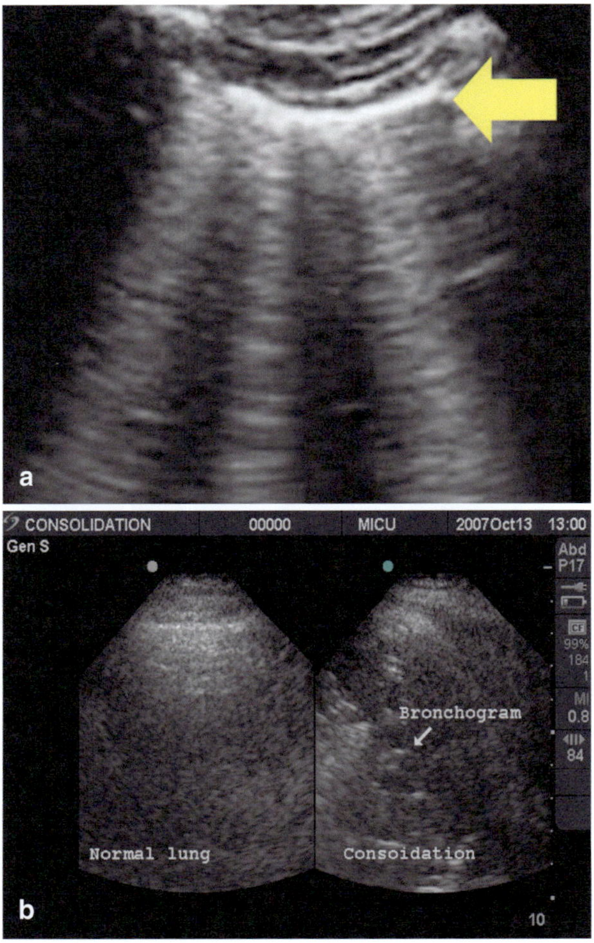

Fig. 6.1 Lung ultrasound for evaluation of lung contusion. The sonographic patterns indicative of lung contusion include "alveolar interstitial syndrome (AIS)" which is identified by an increase in B-lines with no clinical suspicion of cardiogenic pulmonary edema or the presence of a "peripheral parenchymal lesion (PPL)" defined as observation of C-lines (confluent consolidations or "hepatization") which are hypoechoic subpleural focal lesions which identify areas of focal lung consolidation. (**a**) B-lines are vertical densities extending from the pleura (arrow) throughout the entire depth of the scanning field. Three or more B-lines within one lung segment suggest increased interstitial lung water consistent with passive congestion, pulmonary edema, infection, and/or inflammation involving that anatomic area. (**b**) C-lines are hypoechoic areas of confluent consolidation in the subpleural region which identify areas of focal lung consolidation or lung contusion

planning rib fracture fixation (Fig. 6.3). Thoracic CT scan imaging also provides discrimination between retained hemothorax vs. worsening lung contusion, assisting in determination of best treatment strategies to improve pulmonary function such as VATS or rib fracture fixation (Fig. 6.4).

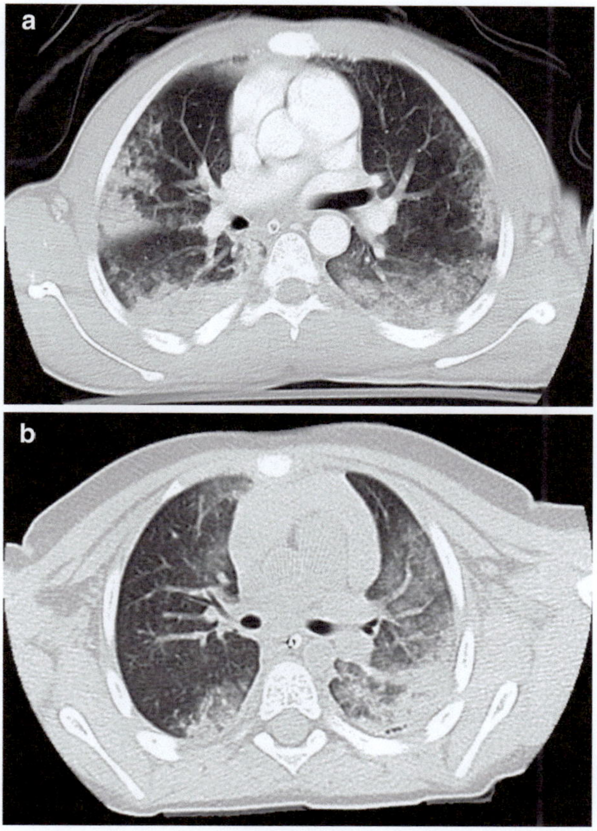

Fig. 6.2 (**a**, **b**) Lung contusions diagnosed by chest CT scan imaging

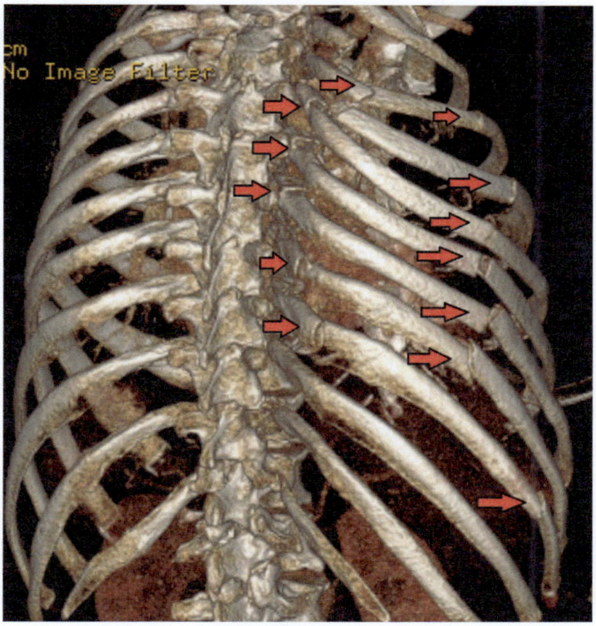

Fig. 6.3 Rib fractures, particularly flail chest, are commonly associated with pulmonary contusion, and thoracic CT scan imaging with bony reformats provides excellent detail of the fracture pattern

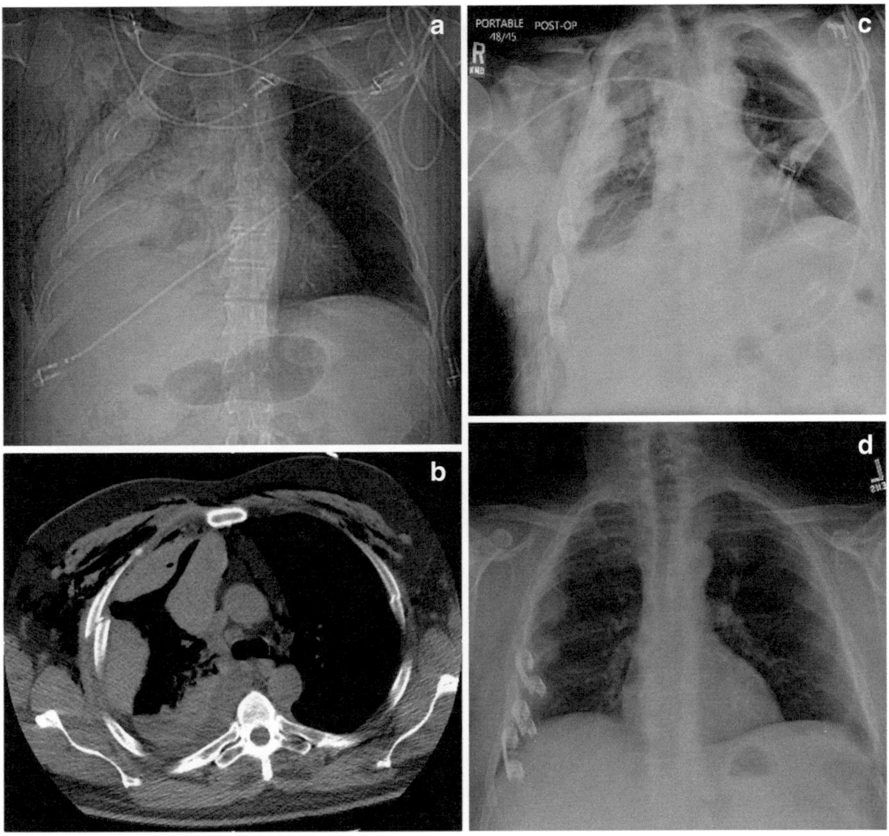

Fig. 6.4 Thoracic CT. A 57-year-old man was involved in fall down 12 stairs landing on his right side. Initial CXR and CT imaging revealed bilateral rib fractures with multiple sequential ribs (2–7) on the right fractured in two locations consistent with flail chest morphology, large volume right hemothorax and pulmonary contusion of most of the right lung and small pneumothorax, pneumomediastinum, and subcutaneous emphysema. Right tube thoracostomy drained 400 cc of blood and minimal thereafter. Thoracic epidural was placed for pain control. He was admitted to the ICU for monitoring and did not require intubation. He had worsening pain and worsening respiratory insufficiency related to unstable right flail chest. (**a**, **b**) This patient underwent VATS/rib fracture fixation which resulted in marked improvement in chest wall pain and prompt weaning of supplemental oxygen and improved pulmonary toilet. CXR post-op and 1 month later are below. (**c**, **d**) The current EAST practice management guideline for pulmonary contusion—flail chest states that surgical fixation may be considered in severe unilateral flail chest or in patients requiring mechanical ventilation when thoracotomy is otherwise required [15]. No large randomized trial data for early rib fracture fixation exist at present, but three small randomized trials and multiple meta-analyses suggest some benefits (decreased pneumonia, ventilator days, and length of stay)

Whole-body CT (WBCT) imaging has emerged as an optimal diagnostic tool in severely injured patients. A recent study from the trauma registry of the German Trauma Society examined a total of 13,564 patients (pre-WBCT, $n = 5005$; WBCT, $n = 8559$) and documented that significantly more clinically relevant thoracic injuries were identified using WBCT, including lung contusions (pre-WBCT, 18.5%; WBCT, 28.7%), injuries to the lung parenchyma (pre-WBCT, 12.6%; WBCT, 5.9%), multiple rib fractures (pre-WBCT, 10.6%; WBCT, 21.6%), and pneumothoraces (pre-WBCT, 17.3%; WBCT, 21.6%) [13].

Some studies have documented that the frequent use of chest CT in blunt trauma may diagnose clinically irrelevant lung contusion [14]. Of 21,382 patients, 8661 (40.5%) had both CXR and chest CT, and 1012 (11.7%) of these had lung contusion, making it the second most common injury after rib fracture. Lung contusion was seen on chest CT only (SOCTO) in 739 (73.0%). Most (73.5%) lung contusion patients had other thoracic injuries. Lung contusion patients had higher admission rates (91.9% vs. 61.7%; mean difference, 30.2%; 95% CI, 28.1–32.1%) and mortality (4.7% vs. 2.0%; mean difference, 2.8%; 95% CI, 1.6–4.3%) than non-lung contusion patients, but mortality was restricted to patients with other injuries (ISS >10). Patients with lung contusion SOCTO had low rates of associated mechanical ventilation (4.6%) and low mortality (2.6%), comparable to that of patients without lung contusion. This study confirmed that lung contusion is commonly diagnosed under current blunt trauma imaging protocols, and most PC are SOCTO with other thoracic injury.

Lung Contusion Management

Optimal treatment of lung contusion due to blunt thoracic trauma is mostly supportive, aiming to prevent acute respiratory failure and maintain adequate oxygenation and ventilation [15]. Frequent physiologic monitoring is required, with meticulous monitoring of vital signs, oxygenation by pulse oximetry, ventilation by arterial or venous blood gases, overall respiratory function by monitoring respiratory rate and incentive spirometry performance, fluid balance by monitoring intake and output, and clinical examination. These patients are best cared for in an intermediate- or intensive care unit given the significant likelihood of clinical deterioration and the possible development of acute respiratory failure. Associated complications such as acute respiratory failure, pneumonia, acute respiratory distress syndrome, and long-term pulmonary disability are frequent sequelae of lung contusion.

Lung Contusion Management: Noninvasive

All lung contusion patients should receive "noninvasive" standard supportive care. This includes pulmonary toilet with incentive spirometry, early mobility, supplemental oxygen, fluid restrictive strategy (minimize intravenous fluids), appropriate pain control (particularly if flail chest and rib fractures), and serial chest radiograph evaluation. Additional noninvasive lung contusion management strategies include the use of noninvasive ventilation and heated high-flow nasal cannula (HFNC) therapy for acute respiratory insufficiency.

Significant advantages of noninvasive ventilation (as compared to tracheal intubation) include decreased risk of ventilator-associated pneumonia, normal speech, swallowing and coughing, and no need for sedation for tracheal tube tolerance. Potential disadvantages of noninvasive ventilation include possible aspiration risk, skin damage, eye irritation, and interruption of diet and secretion clearance.

A recent advance in noninvasive ventilation is high-flow nasal cannula (HFNC) rather than noninvasive positive-pressure ventilation (NIPPV) with face or nasal CPAP or BiPAP. HFNC is a novel technique of oxygen therapy that delivers heated and humidified oxygen at a rate up to 60 L/min. Although HFNC has not been specifically studied in clinical trials in patients with lung contusions, there are a number of studies in adult patients with acute respiratory failure. A recent systematic review and meta-analysis included 18 trials with a total of 3881 adult patients with acute respiratory failure. Compared with conventional oxygen therapy, HFNC was associated with a lower rate of endotracheal intubation, while no significant difference was found in comparison with NIPPV. Since patient tolerance to HFNC is far superior than NIPPV, HFNC is a more reliable alternative than NIPPV to reduce the rate of endotracheal intubation in patients with acute respiratory failure [16].

Clinical Guidelines and Protocols

The Eastern Association for the Surgery of Trauma practice management guideline for the management of pulmonary contusion and flail chest [17] recommends the use of optimal analgesia, fluid restrictive strategy, and aggressive chest physiotherapy to minimize the likelihood of respiratory failure and need for ventilator support. A trial of noninvasive ventilation should be considered in alert compliant patients with worsening respiratory status, and invasive mechanical ventilation is recommended for the treatment of acute respiratory failure, not solely for the purpose of overcoming chest wall instability.

The Western Trauma Association algorithm for rib fractures [18] does not specifically address lung contusion management, but many rib fractures have associated lung contusion, and the recommended management principles of pain control, pulmonary hygiene, and repeat radiologic imaging are all appropriate for patients with lung contusion.

A recent joint practice management guideline from EAST and the Trauma Anesthesiology Society provided recommendations regarding catheter-based analgesia in adult patients with blunt thoracic trauma [19]. They recommended epidural analgesia over non-regional modalities for pain control, although the quality of evidence was poor.

The Harborview Trauma Center rib fracture management protocol was developed based on national guidelines and recommends initial multimodal systemic analgesia with subsequent consideration of a neuraxial catheter placement if the patient has persistent pain and/or no improvement in respiratory parameters (Fig. 6.5). The pathway uses a PIC scoring tool to serially evaluate and monitor patients, referring to pain, inspiratory capacity, and cough [20].

Lung Contusion Management: Invasive

"Invasive" management strategies required in patients with lung contusion include thoracentesis, placement of tube thoracostomy or 14 French pigtail catheters for drainage of pneumothorax or hemothorax, endotracheal intubation, and initiation of mechanical ventilation. Every attempt should be made to fully drain traumatic hemothorax to clear the pleural space and provide optimal lung inflation, including the potential use of intrapleural tPA and DNAse to facilitate hemothorax clearance from the pleural space [21]. We have increasingly used 14-French pigtail catheters rather than large (32–40 French) chest tubes for hemothorax drainage, as they are associated with markedly decreased pain, facilitate increased pulmonary toilet secondary to decreased pain, and provide equivalent hemothorax drainage [22].

Most lung contusion patients who will require "invasive" management are those who develop acute respiratory failure, severe hypoxemia or hypercarbia, and ARDS. It is challenging to predict which lung contusion patients will progress to develop acute respiratory failure and/or ARDS, and a number of studies have identified potential strategies to identify this high-risk trauma patient population.

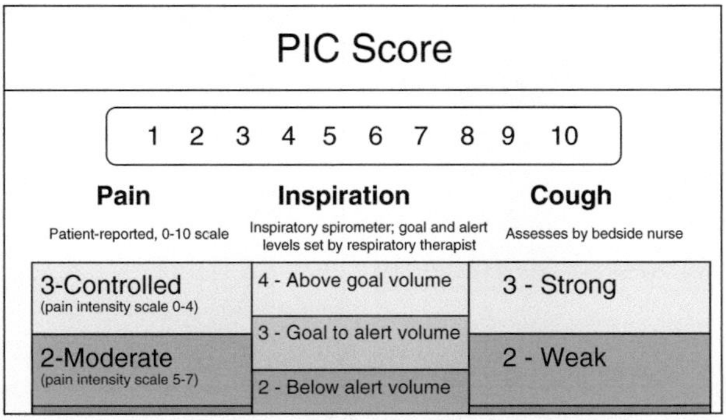

Fig. 6.5 Rib fracture management protocol from Harborview Medical Center. ICU, intensive care unit; IS, incentive spirometry; IV, intravenous; PIC, pain, inspiratory capacity, and cough; PCA, patient-controlled analgesia (From Witt CE and Bulger EM. Comprehensive approach to the management of the patient with multiple rib fractures: a review and introduction of a bundled rib fracture management protocol. Trauma Surg Acute Care Open. 2017;2:1–7, with permission)

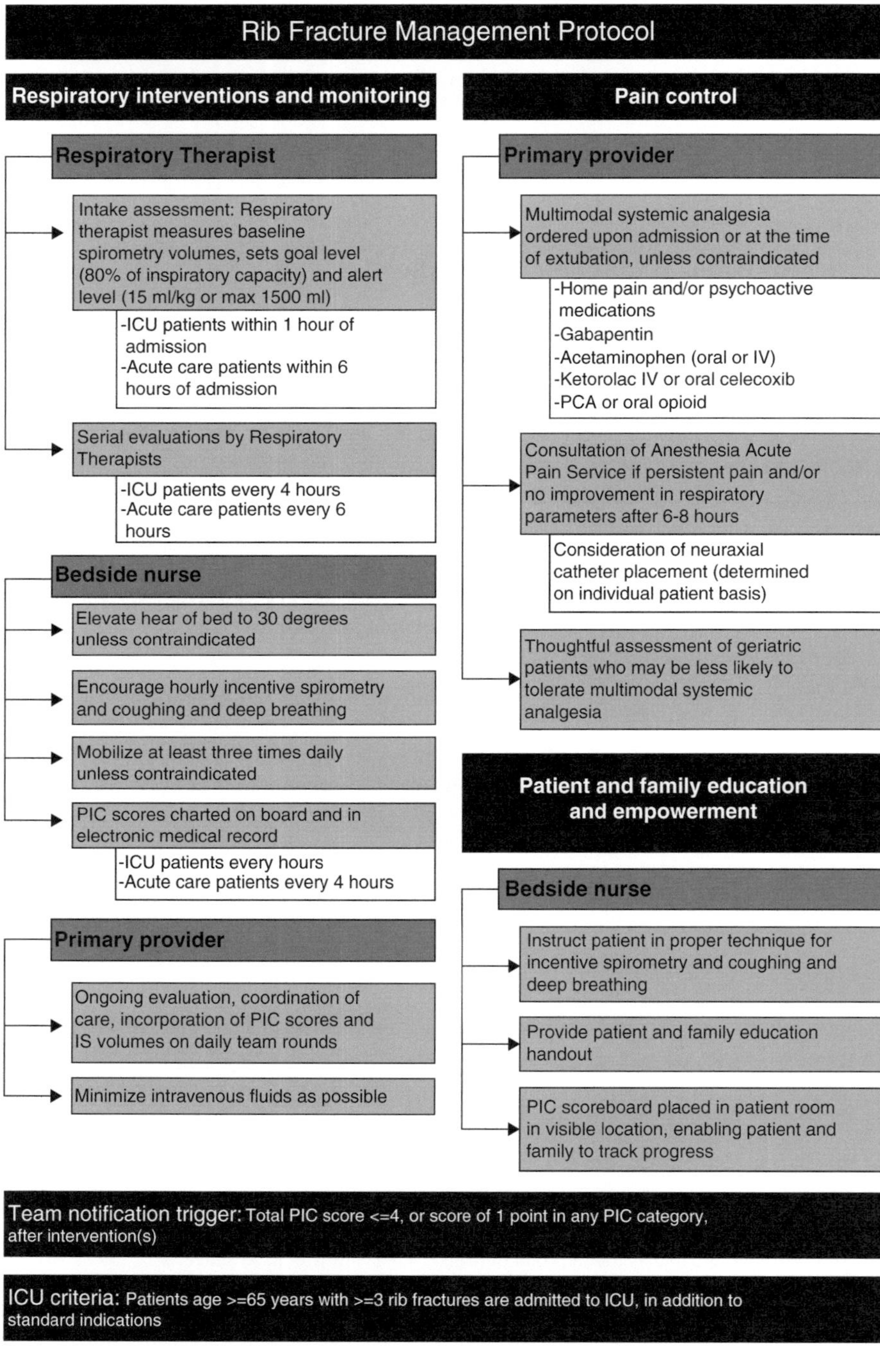

Fig. 6.5 (continued)

Lung Contusion and CT Scan Predictors of Need for Mechanical Ventilation

A single-institution study of 392 patients with blunt thoracic injuries and chest CT imaging on arrival included 243 patients with confirmed lung contusion. Only 25 (6%) of patients required mechanical ventilation. The combination of Glasgow Coma Scale (GCS) score <14, BPC score >2, and >4 ribs fractured predicted mechanical ventilation in 100% of the cases, and the absence of all factors precluded MV in 100%. This algorithm was tested prospectively over 6 months in 55 patients with lung contusion, and they confirmed that the absence of the three factors precluded the need of mechanical ventilation. This study confirmed that a simple score derived by the initial chest CT, in combination with GCS and the number of fractured ribs, can predict the need for mechanical ventilation early after blunt thoracic injury with documented lung contusion [23].

Thoracic Trauma Severity Score on Admission to Predict ARDS

The Thoracic Trauma Severity (TTS) score (Table 6.1) upon admission has been assessed as a predictor of delayed ARDS in blunt trauma patients with lung contusion confirmed by chest CT. Of 329 patients (75% men, mean age 36.9 years [SD 17.8 years], mean Injury Severity Score 21.7 [SD 16.0]), 82 (25%) presented with ARDS (mean lowest PaO2/FiO2 ratio of 131 [SD 34]). The area under the ROC curves for the TTS score in predicting ARDS was 0.82 (95% CI 0.78–0.86). TTS scores a. A TTS score of 13–25 was found to be an independent risk factor for ARDS (OR 25.8 [95% CI, 6.7–99.6] $P < 0.001$) in blunt thoracic trauma [24].

Table 6.1 Thoracic Trauma Severity (TTS) score

PaO2/FiO2	Rib fracture	Contusion	Pleural involvement	Age (yrs)	Points
>400	0	None	None	<30	0
300–400	1–3	1 lobe	Pneumothorax	30–41	1
200–300	4–6 unilateral	1 lobe bilateral or 2 lobes unilateral	Unilateral HT or HPT	42–54	2
150–200	>3 bilateral	<2 lobes bilateral	HT or HPT bilateral	55–70	3
<150	Flail chest	>=2 lobes bilateral	Tension pneumothorax	>70	5

All categories have to be added to achieve a score ranging from 0 to 25
HT hemothorax, *HPT* hemopneumothorax
From Daurat A, Millet I, Roustan JP, Maury C, Taourel P, Jaber S, Capdevila X, Charbit J. Thoracic Trauma Severity score on admission allows to determine the risk of delayed ARDS in trauma patients with pulmonary contusion. Injury. 2016;47(1):147–53, with permission

Measurement of Lung Contusion Volume by CT Imaging to Predict ARDS

It is currently recognized that there is no fully accurate method to quantitate the amount of lung contusion and pulmonary injury. The ability to predict which lung contusion patients are at increased risk of developing complications (pneumonia, acute respiratory failure, ARDS) would be of great clinical utility. Quantification of lung contusion volume by CT scan imaging has been used as a method to identify patients at high risk for ARDS after lung contusion.

A single-institution study of 49 patients with lung contusion used computer-generated three-dimensional reconstruction from admission chest CT scan to measure lung contusion volume expressed as a percentage of total lung volume. The average lung contusion volume was 18% (range, 5–55%). Patients were classified using contusion volume as severe ($\geq$20%, n = 17) or moderate (<20%, n = 32). There was a much higher rate of ARDS in the severe group compared with the moderate group (82% vs. 22%, $P < 0.001$). There was a trend toward higher pneumonia rate in the severe group compared with the moderate group (50% vs 28%, $P = 0.20$). Injury Severity Score was similar in the severe and moderate groups (23.3 vs. 26.5, $P = 0.33$), as were admission Glasgow Coma Scale score (12 vs. 13, $P = 0.30$), admission blood pressure (131 vs. 129 mm Hg, $P = 0.90$), and admission PaO_2/FiO_2 (197 vs. 255, $P = 0.14$). The authors concluded that measurement of lung contusion volume from the admission chest CT scan identified patients at high risk for ARDS [25].

Another similar single-institution study investigated the clinical correlation between lung contusion volume and patient outcome ($n = 226$) in blunt chest trauma. Motor vehicle crash (54.4%) and falls (16.4%) were the most frequent mechanism of injury. Bilateral lung contusion (61.5%) was more prevalent than right (19.5%) and left lung contusion (19%). In contrast to the study above, lung contusion volume had a significant positive correlation with ISS; age and PaO_2/FiO_2 ratio showed a negative correlation ($P < 0.05$). The median lung contusion volume was significantly higher in patients who developed in-hospital complications ($P = 0.02$). A lung contusion volume >20% was associated with increased risk of ARDS, blood transfusion, and prolonged mechanical ventilation. Presence of chest infection, lung contusion volume measured by CT, and ISS were predictors of ARDS. Similar to the previous study, quantification of lung contusion volume by CT imaging could identify patients at high risk of ARDS [26].

A more recent study ($n = 202$ patients with lung contusion on admission CT) measured percent lung contusion volume using a novel semiautomated, attenuation-defined computer-based algorithm, in which the lung was segmented with minimal manual editing. ARDS developed in 75 patients (37.1%) and pneumonia in 26 patients (12.9%). They found that a lung contusion size $\geq$24% of total lung volume was most significant at predicting ARDS, which occurred in 78% of patients. These patients also had a significantly higher incidence of pneumonia when compared with those with contusions less than 24%. The specificity of contusion size of 24% or greater was 94%, although sensitivity was 37%; positive predictive value was 78%, and negative predictive value was 72% [27].

Measurement of Lung Contusion Volume by Lung Ultrasonography to Predict ARDS

Lung contusion volume measurement by lung ultrasonography (LUS) has also been investigated as a method to predict subsequent ARDS [28]. In a prospective study of 45 blunt trauma patients, the diagnostic accuracy of LUS was compared to that of combined clinical examination and chest radiography for pneumothorax, lung contusion, and hemothorax, with thoracic CT scan as reference. The diagnostic criteria for LUS were based on international recommendations for LUS [29]. Among the 41 patients alive at 72 h, 19 (46%) developed ARDS (5 severe, 9 moderate, 5 mild according to the Berlin classification). Lung contusion extent assessed by LUS on admission was predictive of ARDS within 72 h (AUC-ROC = 0.78 [95% CI 0.64–0.92]). The extent of lung contusion on LUS correlated well with CT scan measurements (Spearman's coefficient = 0.82).

A LUS score of 6 out of 16 was the best threshold to predict ARDS, with a 58% [95% CI 36–77] sensitivity and a 96% [95% CI 76–100] specificity. The diagnostic accuracy of LUS was higher than that of combined clinical examination and chest radiography: (AUC-ROC) 0.81 [95% CI 0.50–1.00] vs. 0.74 [0.48–1.00] ($P = 0.24$) for pneumothorax, 0.88 [0.76–1.00] vs. 0.69 [0.47–0.92] ($P < 0.05$) for lung contusion, and 0.84 [0.59–1.00] vs. 0.73 [0.51–0.94] ($P < 0.05$) for hemothorax. This study confirmed that LUS on admission identified patients at risk of developing ARDS after blunt trauma and provided rapid and accurate diagnosis of common traumatic thoracic injuries.

Management of ARDS Due to Lung Contusion

Some patients with lung contusion will progress to have acute respiratory failure and subsequent acute respiratory distress syndrome (ARDS) and severe hypoxemia. For ARDS due to lung contusion, all advanced ARDS treatment strategies (low tidal volume ventilation, restrictive fluid strategy, neuromuscular blockade, prone position, inhaled nitric oxide, and extracorporeal membrane oxygenation, Fig. 6.6) will be considered, depending on the severity of the hypoxemia based on Berlin ARDS criteria [30]. Management of ARDS patients due to lung contusion must be personalized, as there are many factors to consider in the decision-making with regard to which treatment strategies are best for the individual patient. For instance, trauma patients with significant unstable spine injuries cannot be proned. Veno-venous ECMO is not contraindicated in traumatic lung contusion, since newer ECMO circuits can be managed without systemic anticoagulation if high-flow rates are maintained. Lung contusion patients with severe ARDS (PaO_2/FiO_2 ratio <100) should therefore be managed at an ARDS center with ECMO capabilities.

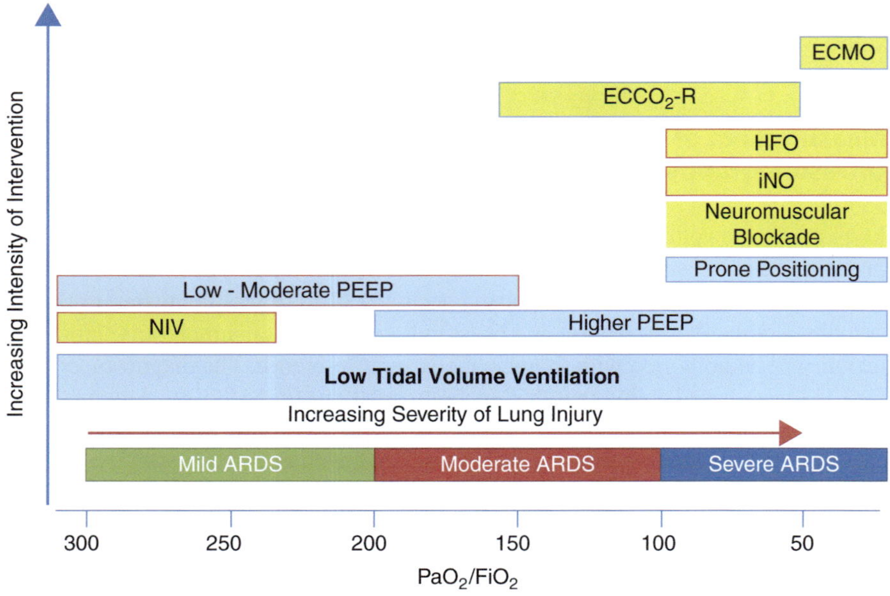

Fig. 6.6 Rescue strategies in ARDS based on severity of hypoxemia by Berlin criteria

References

1. Fallon M. Lung injury in the intact thorax with report of a case. Br J Surg. 1940;28:39–49.
2. Karmy-Jones R, Jurkovich GJ. Blunt chest trauma. Curr Probl Surg. 2004;41(3):211–380.
3. Alfano GS, Hale HW Jr. Pulmonary contusion. J Trauma. 1965;5(5):647–58. PMID: 5826710
4. Demuth WE Jr, Smith JM. Pulmonary contusion. Am J Surg. 1965;109:819–23. PMID: 14283348
5. Stevens E, Templeton AW. Traumatic nonpenetrating lung contusion. Radiology. 1965;85:247–52.
6. Lloyd JW, Smith AC, O'Connor BT. Classification of chest injuries as an aid to treatment. Br Med J. 1965;1(5449):1518–23.
7. Probst C, Pape H-C, Hildebrand F, Regel G, Mahlke L, Giannoudis P, et al. 30 years of polytrauma care: an analysis of the change in strategies and results of 4849 cases treated at a single institution. Injury. 2009;40:77–83.
8. Soldati G, Testa A, Silva FR, Carbone L, Portale G, Silveri NG. Chest ultrasonography in lung contusion. Chest. 2006;130(2):533–8. http://5minsono.com/lung-bruise/
9. http://5minsono.com/lung-bruise/; accessed 11/20/2017.
10. Gargani L, Volpicelli G. How I do it: lung ultrasound. Cardiovasc Ultrasound. 2014;12:25.
11. Trupka A, Waydhas C, Hallfeldt KK, Nast-Kolb D, Pfeifer KJ, Schweiberer L. Value of thoracic computed tomography in the first assessment of severely injured patients with blunt chest trauma: results of a prospective study. J Trauma. 1997;43(3):405–11. discussion 411-2
12. Omert L, Yeaney WW, Protetch J. Efficacy of thoracic computerized tomography in blunt chest trauma. Am Surg. 2001;67(7):660–4.
13. Lang P, Kulla M, Kerwagen F, Lefering R, Friemert B, Palm HG, TraumaRegister DGU. The role of whole-body computed tomography in the diagnosis of thoracic injuries in severely injured patients - a retrospective multi-Centre study based on the trauma registry of the German

trauma society (TraumaRegister DGU®). Scand J Trauma Resusc Emerg Med. 2017;25(1):82. https://doi.org/10.1186/s13049-017-0427-4.
14. Rodriguez RM, Friedman B, Langdorf MI, Baumann BM, Nishijima DK, Hendey GW, Medak AJ, Raja AS, Mower WR. Pulmonary contusion in the pan-scan era. Injury. 2016;47(5):1031–4. https://doi.org/10.1016/j.injury.2015.11.043. Epub 2015 Nov 30
15. Cohn SM, Dubose JJ. Pulmonary contusion: an update on recent advances in clinical management. World J Surg. 2010;34(8):1959–70. https://doi.org/10.1007/s00268-010-0599-9.
16. Ni YN, Luo J, Yu H, Liu D, Ni Z, Cheng J, Liang BM, Liang ZA. Can high-flow nasal cannula reduce the rate of endotracheal intubation in adult patients with acute respiratory failure compared with conventional oxygen therapy and noninvasive positive pressure ventilation?: a systematic review and meta-analysis. Chest. 2017;151(4):764–75.
17. Simon B, Ebert J, Bokhari F, Capella J, Emhoff T, Hayward T 3rd, Rodriguez A, Smith L, Eastern Association for the Surgery of Trauma. Management of pulmonary contusion and flail chest: an eastern Association for the Surgery of trauma practice management guideline. J Trauma Acute Care Surg. 2012;73(5 Suppl 4):S351–61.
18. Brasel KJ, Moore EE, Albrecht RA, de Moya M, Schreiber M, Karm-Jones R, Rowell S, Namias N, Cohen M, Shatz DV, Biffl WL. Western trauma association critical decisions in trauma: management of rib fractures. WTA 2016 algorithm. J Trauma Acute Care Surg. 2017;82:200–3.
19. Galvagno SM Jr, Smith CE, Varon AJ, Hasenboehler EA, Sultan S, Shaefer G, To KB, Fox A, Alley DE, Ditillo M, et al. Pain management for blunt thoracic trauma: a joint practice management guideline from the eastern Association for the Surgery of trauma and trauma anesthesiology society. J Trauma Acute Care Surg. 2016;81:936–51.
20. Witt CE, Bulger EM. Comprehensive approach to the management of the patient with multiple rib fractures: a review and introduction of a bundled rib fracture management protocol. Trauma Surgery & Acute Care Open. 2017;2:1–7.
21. Koppurapu V, Meena N. A review of the management of complex Para-pneumonic effusion in adults. J Thorac Dis. 2017;9(7):2135–41.
22. Bauman ZM, Kulvatunyou N, Joseph B, Jain A, Friese RS, Gries L, O'Keeffe T, Tang AL, Vercruysse G, Rhee P. A prospective study of 7-year experience using percutaneous 14-French pigtail catheters for traumatic hemothorax/Hemopneumothorax at a Level-1 trauma center: size still does not matter. World J Surg. 2017. https://doi.org/10.1007/s00268-017-4168-3. [Epub ahead of print]
23. de Moya MA, Manolakaki D, Chang Y, Amygdalos I, Gao F, Alam HB, Velmahos GC. Blunt pulmonary contusion: admission computed tomography scan predicts mechanical ventilation. J Trauma. 2011;71(6):1543–7.
24. Daurat A, Millet I, Roustan JP, Maury C, Taourel P, Jaber S, Capdevila X, Charbit J. Thoracic trauma severity score on admission allows to determine the risk of delayed ARDS in trauma patients with pulmonary contusion. Injury. 2016;47(1):147–53. https://doi.org/10.1016/j.injury.2015.08.031. PMID: 26358517
25. Miller PR, Croce MA, Bee TK, Qaisi WG, Smith CP, Collins GL, Fabian TC. ARDS after pulmonary contusion: accurate measurement of contusion volume identifies high-risk patients. J Trauma. 2001;51(2):223–8. discussion 229-30
26. Mahmood I, El-Menyar A, Younis B, Ahmed K, Nabir S, Ahmed MN, Al-Yahri O, Mahmood S, Consunji R, Al-Thani H. Clinical significance and prognostic implications of quantifying pulmonary contusion volume in patients with blunt chest trauma. Med Sci Monit. 2017;23:3641–8. PMID: 28746303
27. Becher RD, Colonna AL, Enniss TM, Weaver AA, Crane DK, Martin RS, Mowery NT, Miller PR, Stitzel JD, Hoth JJ. An innovative approach to predict the development of adult respiratory distress syndrome in patients with blunt trauma. J Trauma Acute Care Surg. 2012;73(5):1229–35.
28. Leblanc D, Bouvet C, Degiovanni F, Nedelcu C, Bouhours G, Rineau E, Ridereau-Zins C, Beydon L, Lasocki S. Early lung ultrasonography predicts the occurrence of acute respiratory

distress syndrome in blunt trauma patients. Intensive Care Med. 2014;40(10):1468–74. https://doi.org/10.1007/s00134-014-3382-9.
29. Volpicelli G, Elbarbary M, Blaivas M, Lichtenstein DA, Mathis G, Kirkpatrick AW, Melniker L, Gargani L, Noble VE, Via G, Dean A, Tsung JW, Soldati G, Copetti R, Bouhemad B, Reissig A, Agricola E, Rouby JJ, Arbelot C, Liteplo A, Sargsyan A, Silva F, Hoppmann R, Breitkreutz R, Seibel A, Neri L, Storti E, Petrovic T. International evidence-based recommendations for point-of-care lung ultrasound. Intensive Care Med. 2012;38:577–91.
30. ARDS Definition Task Force, Ranieri VM, Rubenfeld GD, Thompson BT, Ferguson ND, Caldwell E, Fan E, Camporota L, Slutsky AS. Acute respiratory distress syndrome: the berlin definition. JAMA. 2012;307(23):2526–33.

Rib Fracture and Lung Contusion: Impact on Pulmonary Function Tests

7

Abraham Sonny and Richard M. Pino

Rib fractures and lung contusions are common sequelae of blunt trauma to the chest. An understanding of the pathophysiology of these injuries and factors that may influence pulmonary function is critical to determining treatment and prognosis of these trauma patients.

This chapter will review the immediate and long-term impacts of rib fractures and lung contusions on pulmonary function tests (PFTs). The role of the PFT in risk-stratifying patients presenting with thoracic trauma will be reviewed in detail.

Pathophysiology

Fracture of three or more ribs in two or more places, with or without sternal fracture, leads to creation of a floating segment called a flail chest. This most commonly occurs after a blunt mechanism of injury. Clinically, this is seen as a segment of the chest that moves paradoxically with respiration and when associated with paradoxical movement is called a mechanical flail. If there is no paradoxical movement, this is referred to as a radiographic flail.

In spontaneously breathing patients, the contraction of the diaphragm and the upward and outward movement of the chest wall from intercostal muscle contraction during inspiration generate negative intrapleural pressure necessary for lung expansion. However, a flail segment in the chest wall moves independently based on

A. Sonny (✉) · R. M. Pino
Department of Anesthesia, Critical Care and Pain Medicine, Massachusetts General Hospital, Harvard Medical School, Boston, MA, USA
e-mail: asonny@mgh.harvard.edu; rpino@mgh.harvard.edu

M. de Moya, J. Mayberry (eds.), *Rib Fracture Management*,
https://doi.org/10.1007/978-3-319-91644-6_7

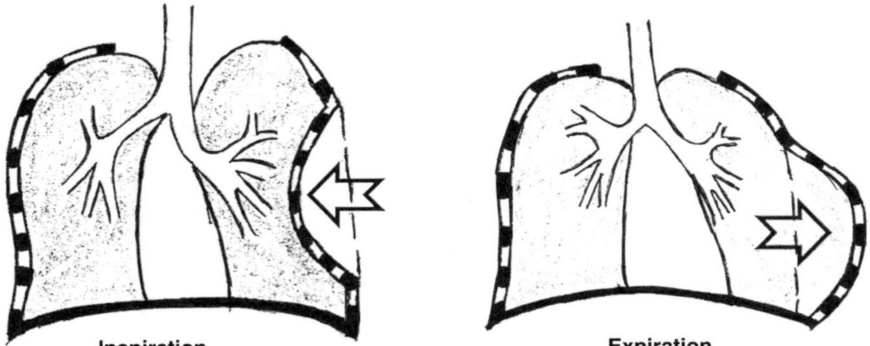

Fig. 7.1 Mechanism of the impact of flail chest on pulmonary function tests

the intrapleural pressure and thus mechanistically plays a counterproductive role during inspiration and expiration (Fig. 7.1). During inspiration when the chest wall expands, the flail segment is pulled inwards, thereby reducing tidal volume (Vt) and vital capacity (VC). Conversely, expiration results in bowing out of the flail segment of the chest wall, creating an area of the lung that cannot completely exhale air. This in turn reduces expiratory volumes and may decrease vital capacity. Such abnormal and ineffective movement of air increases the work of breathing. In addition, pain from rib fractures even in the absence of flail chest may limit the effort generated by the patient and thus can adversely influence the PFT. This may vary depending on the number of ribs fractured and the quality of pain control achieved.

Prior to the 1960s, the paradoxical movement of the chest wall from flail chest was thought to be the primary driver for pulmonary compromise after chest trauma. As a result of this assumption, treatments were aimed mainly at reducing the paradoxical chest wall movement. It has been subsequently shown that the underlying lung contusion, rather than the flail, contributes to a major portion of the immediate respiratory compromise after chest trauma [1]. Lung contusions are direct contusion damage from the chest wall trauma that may include lacerations. This trauma causes the extravasation of plasma and blood into the alveoli that leads to pulmonary edema. This edema reduces lung compliance, increases the shunt fraction (flow past unventilated or collapsed alveoli), and elevates the alveolar-arterial (A-a) gradient.

Lung contusions may also affect the uninjured lung remote from the site of trauma. Porcine models of unilateral blunt chest trauma have shown the presence of non-cardiogenic pulmonary edema in the ipsilateral as well as contralateral non-injured lung a few hours after the initial injury [2, 3]. This is likely a result of delayed capillary leak from progressive secondary inflammatory response [3]. This supports the concept that thoracic trauma may cause varying degrees of pulmonary edema, which in severe cases, may progress to acute respiratory distress syndrome (ARDS). Miller et al. classified the volume of lung contusions as a percentage of the total lung volume using three-dimensional reconstruction from CT scan obtained at hospital admission. They found that among patients with a contusion volume of more than 20%, over three quarter patients developed acute respiratory distress syndrome [4].

Pulmonary edema and atelectasis from chest trauma may cause shunting and/or ventilation/perfusion (V/Q) mismatch with the clinical presentation of hypoxemic respiratory failure. This hypoxemia secondary to V/Q mismatch will improve with an increased concentration of administered oxygen, while shunt hypoxemia will be refractive. The reality is that both processes will occur with lung trauma. Increased pulmonary vascular resistance has been shown to occur in area of lung contusion, which may decrease blood flow to the contused lung causing V/Q mismatch [5]. Other studies have shown that pulmonary vascular resistance is not affected by flail chest itself [6]. Batchinsky et al. used CT scan and multiple inert gas elimination techniques to evaluate ventilation and perfusion in different lung regions in a porcine model of blunt thoracic injury. Though V/Q mismatch had some contribution to hypoxemia after lung contusion, their study showed that true shunt was the major contributor [7].

Acute Impact

Mechanistically, lung contusion and flail chest would be expected to produce a restrictive defect in a PFT, but literature on the immediate impact of thoracic trauma on pulmonary function is sparse. Studies evaluating PFTs immediately after thoracic trauma are fraught with various limitations and challenges. A prerequisite for any reliable and complete PFT is the patient's ability to cooperate and give a maximal effort. This is challenging, especially in severely injured patients, since they are often sedated and mechanically ventilated. Concomitant injuries such as head, abdominal, orthopedic, and burn trauma will markedly affect attempts at pulmonary evaluation. The following section elaborates on literature describing PFTs in the acute phase of thoracic trauma in the context of these limitations.

Investigators have shown a marked reduction in VC shortly after injury from thoracic trauma [8–10]. Bakhos et al. evaluated 38 elderly patients (more than 65 years of age) within 48 h of thoracic trauma and found that the mean VC was reduced to 40% of predicted [8]. Kishikawa et al. performed a small, but elegantly designed, prospective study on 18 patients to elucidate the individual impacts of pulmonary contusion and flail chest on PFTs [10]. They performed the analyses after dividing patients into groups based on presence or absence of pulmonary contusions and flail chest. Vital capacity at 2 weeks after injury was reduced to 50–60% of predicted. A key finding was that the reduction in VC was not independently influenced by lung contusion or flail chest.

Inadequate pain control from rib fractures and paradoxical movements of the chest wall from flail segments could also lead to reduced VC. Pulmonary edema and atelectasis resulting from lung contusions reduce lung compliance. This results in a restrictive defect that in turn reduces VC. Since most studies evaluating chest trauma exclude patients who had very severe injury requiring mechanical ventilation, the severity of VC reduction is likely underreported in the literature.

A reduction in functional residual capacity (FRC) is typically seen in patients with restrictive lung disease, in contrast to an obstructive defect that increases FRC. Kishikawa et al. used a multi-breath nitrogen washout method to measure FRC 2 weeks after thoracic trauma [10]. Their study showed an overall reduction in functional residual capacity to 50–60% of predicted. Patients with pulmonary contusions had more severe reductions compared to those who did not. It is likely that atelectasis and pulmonary edema from pulmonary contusion markedly reduces residual volume and thereby impacts FRC. These investigators investigated the presence of an obstructive defect by measuring the forced expiratory volume in 1 s (FEV1) at 2 weeks from injury. The mean FEV1 was only mildly decreased (80% of predicted), compare to the marked reductions in functional residual capacity and vital capacity.

Risk Stratification Based on Pulmonary Function Tests

In addition to investigating the pathophysiology of thoracic trauma, PFTs have also been used to quantify the severity of injury and to predict adverse outcomes. Because comprehensive pulmonary function testing is cumbersome to perform soon after trauma, most investigators have used tests that can be easily performed by a clinician or a respiratory therapist at bedside.

Carver et al. performed the largest study evaluating over 600 patients presenting with thoracic trauma. VC was measured at bedside within 48 h of presentation using a Wright Mark 8 respirometer (nSpire Health, Inc., Longmont, CO). They defined pulmonary complications as need for admission to an intensive care unit, intubation, pneumonia, dependence on home oxygen therapy, or hospital readmission due to pulmonary causes. They reported a 36% increase in pulmonary complications with each 10% decrease in VC. Importantly, a VC of less than 30% of predicted values stood out as a strong independent predictor of pulmonary complications regardless of adjustment for severity of injury, age, the number of rib fractures, and other comorbidities [9].

Bakhos and his coworkers evaluated 38 elderly patients with blunt chest trauma within 48 h of presentation and found that both decreased VC and the percentage of predicted VC were strongly associated with increased length of hospital stay. VC also correlated with discharge to a nonhome location [8]. Thus, VC easily performed at bedside may be used as tool for early risk stratification and can assist with the better triage and closer monitoring of patients who are likely to decompensate. Additional factors that are independently associated with adverse outcomes are age, injury severity score, number of rib fractures, dependence on mechanical ventilation, and a history of chronic obstructive pulmonary disease [8, 9].

Other bedside tests that have been evaluated for risk stratification of patients with chest trauma are incentive spirometry volume and peak expiratory flow rate. Incentive spirometry volume has been shown to correlate strongly with VC [11], while peak expiratory flow rate is affected by an obstructive defect.

Incentive spirometers are widely available on surgical floors and are less expensive than respirometers. In a prospective study of 99 blunt thoracic trauma patients, Butts et al. found that incentive spirometry volumes at admission were associated with need for mechanical ventilation during hospitalization [12]. Two third of patients with an incentive spirometer volume of less than 1 l needed either invasive or noninvasive mechanical ventilation. Peak expiratory flow rate values were not associated with respiratory failure [12]. This is not surprising, considering the absence of any obstructive defect in pulmonary functions tests after thoracic trauma [10].

Long-Term Impact

Understanding of the impact of blunt thoracic trauma on pulmonary function tests will allow for assessment of the potential for long-term disability as well as monitoring interventions that may improve functional outcome. Some investigations have evaluated recovery in the first few months after injury, while others have evaluated recovery years after injury.

The marked reductions in PFT volumes observed immediately after thoracic trauma persist for the first few weeks after hospital discharge. Livingston et al. prospectively evaluated 28 blunt trauma patients at 2 weeks from hospital discharge and found that the VC remained at 40–50% of predicted values. Subsequently, these patients were followed at 3–6-month intervals for many years. The PFTs improved remarkably over the initial 4–6 months after injury for many patients, while recovery was slower for others. The VC improved to 65–90% of predicted values at 4 months after injury for most patients [13].

Kishikawa et al. attempted to identify factors that are associated with slower recovery and longer-term reduction in PFTs. They followed a small group of 18 patients with blunt thoracic injury for 6 months after injury by performing pulmonary function tests five times during this period. Fig. 7.2 shows the differences in recovery of FRC between groups, based on the presence of radiographically identified pulmonary contusion and/or flail chest. They found that FRC recovered to near normal volumes in patients without pulmonary contusions, irrespective of chest wall deformity from flail chest. On the other hand, patients with pulmonary contusions continued to have a reduced FRC at 6 months after injury. Fibrous changes in the contused lung regions were visible on computerized tomography scans performed months to years after injury [10].

Recovery of lung function measured with PFTs after many years following thoracic trauma has been investigated by several investigators. Livingston et al. did PFTs every 3–6 months after injury until the values stabilized. Normalization of all parameters was observed within 12–24 months after injury. This finding was also supported by Amital et al. who evaluated 13 patients who survived severe chest trauma, defined by occurrence of lung contusions, multiple rib fractures, and hemopneumothorax, 1–4 years after injury. Comprehensive pulmonary function testing was performed on 13 patients 1–4 years after injury. FEV1, VC, total lung capacity, and residual volume were above 80% of normal at the time of testing. As expected,

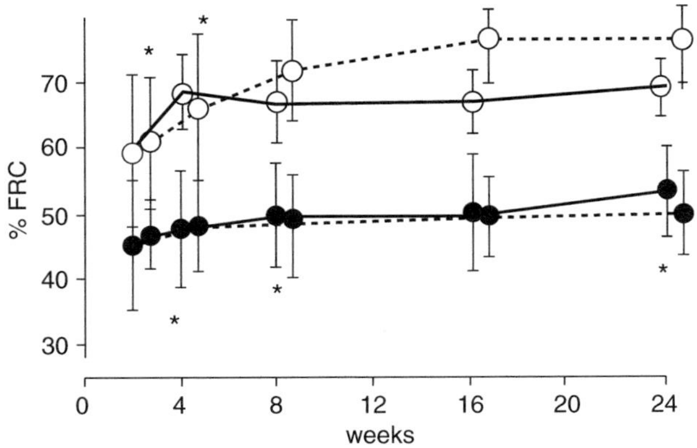

Fig. 7.2 This represents the percentage of predicted functional residual capacity from 2 weeks to 6 months after severe thoracic trauma. Closed circle and solid line denote patients with pulmonary contusion and flail chest (PC+ FC+). Similarly, closed circle and dashed line mean PC+ FC–; open circle and solid line mean PC– FC+; and open circle and dashed line mean PC– FC–. Despite the presence or absence of flail chest, patients with pulmonary contusion (PC+ FC+ and PC+ FC–) have a lower FRC at 6 months than those without pulmonary contusion (PC– FC+ and PC– FC–). FRC, functional residual capacity [10] (Adapted from Kishikawa M et al. Pulmonary contusion causes long-term respiratory dysfunction with decreased functional residual capacity. J Trauma. 1991;31:1203–1208, with permission)

FEV1 was lower in smokers. There was a trend of lower FEV1 among patients who required mechanical ventilation after trauma. Carbon monoxide diffusion capacity was measured to evaluate the effect of pulmonary contusion and resulting fibrosis on diffusion capacity. The authors reported that the diffusion capacity was above 80% of normal at the time of testing that suggested substantial physiological recovery among these patients [14].

Svennevig et al. looked at 24 patients with severe lung contusions and multiple rib fractures 5 years from injury. They measured VC, FEV1, functional residual capacity, total lung capacity, residual volume, and carbon monoxide diffusion that were found to be more than 80% of predicted in most patients. A trend of poorer function, measured as a lower maximal minute ventilation, was found for patients who required mechanical ventilation after trauma [15]. A caveat of this study, however, is that it was performed prior to the standard practice of low tidal volume ventilation as prevention for acute lung injury.

Patients with severe thoracic injury report respiratory symptoms, including dyspnea, years after injury [16]. These subjective symptoms appear to be unrelated to objective results of pulmonary function tests [15]. Table 7.1 summarizes current data on recovery of pulmonary function tests after severe thoracic trauma. Although data suggest that permanent long-term pulmonary disability after thoracic trauma is rare, the results need to be interpreted in context. These data cannot be extrapolated to elderly trauma patients who have been markedly underrepresented in studies. For example, the patient populations studied by Amital et al. and Svennevig et al. had a mean age of 45 and 48 years, respectively [14, 15]. It is also unclear whether preexisting lung disease affects long-term recovery after thoracic trauma.

Table 7.1 Timeline of recovery in pulmonary function tests after severe blunt chest trauma

Timeline	At admission	2 weeks after hospital discharge	4–6 months	>1 year
Vital capacity	50–60%	40–50%	65–90%	>80%
Functional residual capacity	50–60%		50–70%	>80%
FEV1	80%	50–80%		>80%
DLCO				>80%

DLCO diffusing capacity of the lungs for carbon monoxide, *FEV1* forced expiratory volume in 1 s
References [10, 13–15]

Conclusion

Severe thoracic trauma leads to marked impairment in pulmonary function that results in a predominantly restrictive defect. In addition to providing insight into the pathophysiology of thoracic trauma, bedside pulmonary function tests may serve as valuable tools for risk stratification. Though most surviving patients achieve near-complete recovery in pulmonary function tests, the impact of pre-existing pulmonary compromise on recovery is less known.

References

1. Reid JM, Baird WL. Crushed chest injury: some physiological disturbances and their correction. Br Med J. 1965;1:1105–9.
2. Hellinger A, Konerding MA, Malkusch W, Obertacke U, Redl H, Bruch J, et al. Does lung contusion affect both the traumatized and the noninjured lung parenchyma? A morphological and morphometric study in the pig. J Trauma. 1995;39:712–9.
3. Davis KA, Fabian TC, Croce MA, Proctor KG. Prostanoids: early mediators in the secondary injury that develops after unilateral pulmonary contusion. J Trauma. 1999;46:824–31. discussion 831-822
4. Miller PR, Croce MA, Bee TK, Qaisi WG, Smith CP, Collins GL, et al. Ards after pulmonary contusion: accurate measurement of contusion volume identifies high-risk patients. J Trauma. 2001;51:223–8. discussion 229–230
5. Fulton RL, Peter ET. Physiologic effects of fluid therapy after pulmonary contusion. Am J Surg. 1973;126:773–7.
6. Craven KD, Oppenheimer L, Wood LD. Effects of contusion and flail chest on pulmonary perfusion and oxygen exchange. J Appl Physiol Respir Environ Exerc Physiol. 1979;47: 729–37.
7. Batchinsky AI, Jordan BS, Necsoiu C, Dubick MA, Cancio LC. Dynamic changes in shunt and ventilation-perfusion mismatch following experimental pulmonary contusion. Shock. 2010;33:419–25.
8. Bakhos C, O'Connor J, Kyriakides T, Abou-Nukta F, Bonadies J. Vital capacity as a predictor of outcome in elderly patients with rib fractures. J Trauma. 2006;61:131–4.
9. Carver TW, Milia DJ, Somberg C, Brasel K, Paul J. Vital capacity helps predict pulmonary complications after rib fractures. J Trauma Acute Care Surg. 2015;79:413–6.
10. Kishikawa M, Yoshioka T, Shimazu T, Sugimoto H, Yoshioka T, Sugimoto T. Pulmonary contusion causes long-term respiratory dysfunction with decreased functional residual capacity. J Trauma. 1991;31:1203–8. discussion 1208–1210

11. Pinheiro AC, Novais MC, Neto MG, Rodrigues MV, de Souza Rodrigues E Jr, Aras R Jr, et al. Estimation of lung vital capacity before and after coronary artery bypass grafting surgery: a comparison of incentive spirometer and ventilometry. J Cardiothorac Surg. 2011;6:70.
12. Butts CA, Brady JJ 3rd, Wilhelm S, Castor L, Sherwood A, McCall A, et al. Do simple beside lung function tests predict morbidity after rib fractures? Am J Surg. 2017;213:473–7.
13. Livingston DH, Richardson JD. Pulmonary disability after severe blunt chest trauma. J Trauma. 1990;30:562–566; discussion 566-567.
14. Amital A, Shitrit D, Fox BD, Raviv Y, Fuks L, Terner I, et al. Long-term pulmonary function after recovery from pulmonary contusion due to blunt chest trauma. Isr Med Assoc J. 2009;11:673–6.
15. Svennevig JL, Vaage J, Westheim A, Hafsahl G, Refsum HE. Late sequelae of lung contusion. Injury. 1989;20:253–6.
16. Beal SL, Oreskovich MR. Long-term disability associated with flail chest injury. Am J Surg. 1985;150:324–6.

Indications for Rib Fixation

8

Tiffany Zens, Krista Haines, and Suresh Agarwal

Guidelines for Rib Fixation

Despite the resurgence of research supporting operative rib fixation and the evolving technology in this field, surgeons are still underutilizing rib fixation in patients with severe thoracic trauma [1]. One reason for this is that there are currently no absolute or universally accepted indications for operative rib fixation and there is significant variation in the quality of evidence for its relative indications [2]. The primary aims of operative rib fixation are to decrease the duration and intensity of respiratory support by improving pulmonary mechanics, reducing pain, and preventing restrictive lung pathology associated with severe chest wall deformity [3]. Critics argue that the current literature on the benefits of operative rib fixation is based on three small randomized controlled trials (RCTs), a few prospective studies, and several retrospective studies. Additionally, there is concern that long-term outcomes after rib fixation are unknown [4]. Finally, it is difficult to differentiate the outcomes of patients with specific rib fracture patterns given no studies include only non-flail chest patients or subgroup analyses [5].

Recently, the Eastern Association for the Surgery of Trauma (EAST) issued practice guidelines clarifying indications for rib fixation in thoracic trauma patients

T. Zens
Department of Surgery, UW Hospital and Clinics, University of Wisconsin, Madison, WI, USA
e-mail: tzens@uwhealth.org

K. Haines · S. Agarwal (✉)
Division of Trauma and Critical Care Surgery, Department of Surgery, Duke University Medical Center, Durham, NC, USA
e-mail: krista.haines@duke.edu; Suresh.Agarwal@duke.edu

M. de Moya, J. Mayberry (eds.), *Rib Fracture Management*,
https://doi.org/10.1007/978-3-319-91644-6_8

Table 8.1 Indications for rib fixation

Level of support	Indications
Strong evidence to support indication • Consistent with practice guidelines by EAST • Supported by results from small RCT • Meta-analysis results demonstrating benefit	• Flail chest in a patient on mechanical ventilation with no underlying pulmonary contusion
Moderate evidence to support indication • Consistent with expert opinion • Supported by results from small RCTs, retrospective studies, case series, or case-control studies	• Chest wall deformity • Symptomatic rib fracture nonunion • Acute pulmonary herniation • Failure to wean from mechanical ventilation in a patient with severe chest wall trauma • Poor pulmonary mechanics in a patient with severe chest wall trauma • Acute pain control • Improved chronic pain or reduced long-term disability • Thoracotomy for other indications
Poor evidence to support indication • Indication proposed in the literature by individual authors • Supported by results from case reports • Theoretical benefit	• Open rib fracture • Rib fracture number threshold • Severely displaced rib fracture Patient age

[5]. The EAST guidelines state "In adult patients with flail chest after blunt trauma, we conditionally recommend rib open reduction and internal fixation to decrease mortality, shorten duration of mechanical ventilation, intensive care unit (ICU) length of stay (LOS) and hospital LOS, incidence of pneumonia and need for tracheostomy. We cannot offer a recommendation for pain control with currently available evidence. In adult patients with non-flail rib fractures after blunt trauma, we cannot offer a recommendation for any of the outcomes with currently available evidence" [5]. The primary aim of this chapter is to evaluate indications for operative rib fixation and the quality of evidence which supports these recommendations. Table 8.1 summarizes the indications for operative rib fixation discussed in the literature and the level of evidence to support each recommendation.

Indications for Rib Fixation: Anatomical Considerations

Flail Chest

The most common and well-supported indication for operative rib fixation is the presence of a flail chest segment [6]. Operative intervention is more widely accepted in patients with flail chest requiring mechanical ventilation, and a recent survey of 405 surgeons found only 8% believed flail chest not requiring mechanical ventilation was an indication for rib fixation [7]. Flail chest is present in approximately 6% of patients with thoracic trauma [8–10]. Flail chest is defined as an incompetent chest wall segment comprising of three or more ribs fractured in two or more places [11]. The paradoxical wall motion associated with flail chest results in low tidal volumes, alveolar collapse, and arteriovenous shunting and can result in complications including pneumonia, prolonged mechanical ventilation, and death [12].

In patients with flail chest, operative rib fixation is supported by the literature and EAST guidelines [5, 13]. These recommendations resulted primarily from three small RCTs and several meta-analyses of existing data. The first RCT by Tanaka et al. included 37 patients and demonstrated flail chest patients treated with operative fixation required fewer days of mechanical ventilation ($p < 0.05$) and had lower rates of pneumonia ($p < 0.05$), fewer days in the ICU ($p < 0.05$), and lower medical costs [12]. The second RCT by Granetzny et al. of 40 patients with flail chest showed those treated with operative fixation required fewer days of mechanical ventilation ($p < 0.001$) and had decreased ICU LOS ($p < 0.001$), lower incidence of pneumonia ($p = 0.014$), and decreased hospital LOS ($p < 0.001$) [14]. Finally, a RCT by Marasco et al. of 46 flail chest patients concluded those treated with operative fixation had fewer days on mechanical ventilation ($p = 0.01$) and shorter ICU LOS ($p = 0.03$) [15]. In a Cochrane review of these three studies, pooled data showed no evidence of improved mortality with operative fixation (relative risk(RR) 0.56, confidence interval (CI) 0.13–2.42) but decreased rates of pneumonia (RR 0.36, CI 0.15–0.85), chest deformity (RR 0.13, CI 0.03–0.67), and tracheostomy placement (RR 0.38, CI 0.14–1.02) [16]. Additionally, several retrospective studies have demonstrated the benefits of operative fixation in the flail chest population, and as a result, this has become the most prominent indication described in the literature [4, 17–21]. Given the extensive data on improved patient outcomes in this population, several meta-analyses have pooled prospective and retrospective study results and demonstrated rib fixation leads to reduced duration of mechanical ventilation, decreased ICU LOS, decreased hospital LOS, decreased hospital costs, decreased rates of pneumonia, and reduced need for tracheostomy [3, 5, 22–24]. Three of the five meta-analyses also demonstrated improved mortality [3, 5, 24].

Chest Wall Deformity

The presence of significant chest wall deformity is cited by a number of authors in the literature as a relative indication for rib fixation [17, 18, 21, 25, 26] and is accepted by 58% of surveyed surgeons as an appropriate indication [7]. Chest wall deformity results when severely displaced rib fractures result in a loss of thoracic volume from an incompetent or caved in segment of the rib cage. Chest wall deformities impede normal lung expansion [27, 28]. The severely displaced ribs causing chest wall deformity also have the potential to cause pulmonary lacerations, pneumothoraces, or pulmonary hernias [28]. Although there is no universal agreement regarding what severity of chest wall deformity warrants operative intervention, a loss of 30% of the chest wall volume has been proposed by at least one author [20]. The primary data supporting use of chest wall deformity as an indication for rib fixation comes from the RCTs of Marasco et al. and Granetzny et al. which noted decreased chest wall deformity in patients treated operatively [14–16].

Symptomatic Rib Fracture Nonunion or Chronic Malunion

A small segment of patients with severe rib fractures can develop symptomatic nonunion or chronic malunion. Rib fractures that have not previously been fixated can develop a sensation of "clicking" and may affect adjacent structures. Snapping scapula syndrome is one situation that has been described, and case reports have shown delayed rib fracture repair can result in improvement in symptomatology [29, 30]. Symptomatic nonunion occurs when the fractured rib segments fail to heal adequately and result in pseudarthrosis and chronic pain [27, 28, 31–34]. Symptomatic rib fracture nonunion has been proposed by many as a relative indication for operative rib fixation in cases where cross-sectional imaging demonstrates nonunion 2 months after injury and the patient is persistently symptomatic [17, 18, 21, 27, 28, 31–34]. Although the evidence to support this practice is limited to encouraging case series and case reports, 26% of surgeons in the Mayberry et al. survey consider fibrous nonunion an indication for operative rib fixation [7, 31–33]. Unfortunately, no randomized trials or prospective observations of these patients are currently available.

Open Rib Fracture

Although there is currently no published data supporting the use of rib fixation in cases of open fractures, several authors consider this a relative indication. They site standard principals of open fracture management as the rationale for this recommendation [2, 4]. It is important to note in the cases of open rib fractures, absorbable plates and absorbable suture cerclage may be a more appropriate hardware for operative fixation given concerns of contamination in the wound bed to minimize the risk of osteomyelitis and infected hardware [4].

Acute Pulmonary Herniation

An acute pulmonary hernia is an uncommon complication of severe blunt thoracic trauma in which the lung herniates through a defect in the chest wall. Given the possibility of acute complications including pulmonary incarceration or strangulation, operative intervention is often the treatment modality of choice, although conservative therapy has been described in the literature [35]. Many experts consider acute pulmonary herniation a relative indication for rib fixation [4, 18, 25]. The intercostal muscle defects can be closed by suturing the surrounding ribs together and repairing fractures [4]. When surveyed by Mayberry et al., 58% of surgeons cited pulmonary herniation is a valid indication for rib fixation [7].

Number of Rib Fractures or Severely Displaced Fractures

It has been well documented in the literature that there is a direct relationship between the number of rib fractures and increased patient morbidity and mortality

[36]. In fact, a study by Flagel et al. noted a 10% mortality in patients with more than four rib fractures and a 34% mortality in patients with eight or more rib fractures [37]. Furthermore, the impact of multiple rib fractures on morbidity and mortality has been found to be even more pronounced older trauma patients [37–41]. Severely displaced rib fractures have been associated with higher rates of pulmonary complications [42]. As a result of these findings, several authors have proposed the number of rib fractures, or the degree of fracture displacement could be a relative indication for operative rib fixation [6, 12, 20, 43]. Threshold limits of ≥3 severely displaced fractures [20], more than four fractures [6], or more than six fractures [12] have been proposed. Unfortunately, there is no evidence in the literature to support using fracture number or degree of displacement as a sole indication for operative rib fixation. In the Mayberry et al. survey, 27% of respondents believed a rib fracture with displacement ≥1 rib width was an indication for fixation [7].

Indications for Rib Fixation: Physiologic Considerations

Failure to Wean from the Ventilator

In addition to anatomical considerations, several authors have suggested physiologic indications for operative rib fixation in patients with severe thoracic trauma. The most common of these physiologic indications is failure to wean from mechanical ventilation [17, 20, 25, 43–45]. Unfortunately there is currently no consensus regarding at what point operative rib fixation is indicated in order to help liberate a patient from mechanical ventilation. In the Tanaka et al. RCT, patients were enrolled after 5 days of mechanical ventilation [12]. In contrast, the Marasco et al.'s and Granetzny et al.'s RCTs enrolled patients within the first 48 h [14, 15]. In the Mayberry et al. survey, 34% of respondents believed operative fixation was warranted in patients with flail chest and inability to wean from mechanical ventilation after 7 days, while 29% reported operative rib fixation should be considered in patients with flailed chest unable to wean from mechanical ventilation after 14 days [7]. The data to support failure to wean from mechanical ventilation as an indication for operative rib fixation comes from a number of studies finding a decreased number of days on mechanical ventilation and decreased need for tracheostomy in patients undergoing operative fixation when compared to those treated conservatively for rib fractures [12, 17, 43, 44]. A study by Doben et al. demonstrated those patients treated with operative rib fixation were liberalized from mechanical ventilation in a median of 1.5 days [45]. In addition, the two of the three RCT studies demonstrated fewer days on mechanical ventilation in those patients undergoing operative fixation [12, 14].

Poor Pulmonary Mechanics

In addition, several authors have proposed that poor pulmonary mechanics as measured by pulmonary function tests (PFTs) may be an indication for operative rib fixation [12, 14, 22]. Unfortunately, given the feasibility of preforming PFTs on a critically ill, mechanically ventilated patient, there is no consensus for exact PFT thresholds for operative intervention. The data to support this practice comes from several studies showing improved PFTs in patients undergoing operative rib fixation, particularly a decreased restrictive pattern as measured by forced vital capacity and total lung capacity [14, 22]. Additionally, the RCTs by Tanaka et al. and Granetzny et al. demonstrated improved spirometry in the operatively treated patients with flail chest, although the Marasco et al. RCT failed to show this finding [12, 14, 15].

Indications for Rib Fixation: Quality of Life

Pain Control

Acute pain management is one of the hallmarks of the treatment of rib fractures. Chest wall pain is often controlled by multimodal therapies including regional pain control with epidural or paravertebral catheters, ketamine, nonsteroidal anti-inflammatory drugs, acetaminophen, topical pain patches, transcutaneous electrical nerve stimulators, and opiates [46]. Pain from rib fractures leads to splinting of intercostal muscles and atelectasis. In addition, a study by Fabricant et al. found that one of the most predictive factors for chronic pain after rib fractures is the intensity of acute pain in the post-injury period [47]. As a result, several authors have proposed that unrelenting pain refractory to maximum medical management is a relative indication for operative rib fixation [12, 17, 18, 20, 21, 25, 26, 28, 31, 43, 44]. In fact, 10% of academic trauma, orthopedic, and thoracic surgeons surveyed by Mayberry et al. considered persistent poorly controlled acute pain after 7–10 days a valid indication for operative rib fixation [7]. There is conflicting data to support this indication. Work by Khandelwal et al. and De Moya et al. shows decreased pain scores and narcotic requirements in patients after operative rib fixation [48, 49], but several other studies do not corroborate these findings [15, 44].

Reduced Long-Term Pain and Disability

Several studies have investigated the long-term pain and disability associated with severe thoracic trauma. The literature has found that 50–60% of patients with rib fractures develop some long-term morbidity, and 43% will not return to full-time employment. Furthermore, patients lose an average of 70 days of work during their recovery [4, 50–52]. Patients with rib fractures have been found to be significantly more disabled ($p < 0.001$) at 30 days post-injury than patients with chronic medical illnesses [50]. Chronic symptoms in patients after flail chest include long-term dyspnea and chest pain with abnormal PFTs [50].

As a result of these findings, several experts have recommended chronic pain and disability as an indication for operative rib fixation [4, 12, 25, 26, 50, 51, 53]. In fact, the Mayberry et al. survey demonstrated 26% of respondents stated chronic pain was an appropriate indication for rib fixation [7]. The data to support this indication is mixed. Tanaka et al. found flail chest patients who underwent operative fixation were more likely to return to work within 6 months ($p < 0.05$) [12]. Furthermore, Mayberry et al. and Campbell et al. demonstrated patients had low long-term morbidity and pain following operative rib fixation [25, 26]. In contrast, Marasco et al. documented no differences in quality of life, long-term spirometry, or activity levels between those who underwent rib fixation and those who were treated conservatively [15].

Indications for Rib Fixation: Other

Thoracotomy for Other Indications

There are several circumstances in which a thoracotomy is indicated after thoracic trauma including evacuation of a hematoma or repair of a pulmonary injury. When the patient is otherwise stable and a thoracotomy is indicated for another reason, many experts believe it is appropriate to also preform rib fixation [2, 17, 20, 21, 25, 26]. It is thought that by stabilizing rib fractures after completion of the primary surgery, the patient may have improved pain control and pulmonary mechanics [2]. In the survey by Mayberry et al., 18% of academic trauma, orthopedic, and thoracic surgeons surveyed believed rib fixation is warranted after a thoracotomy for other indications [7]. That being said, there is limited data in the literature exploring outcomes for patients who did and did not receive operative rib fixation for this indication.

Age

The final relative indication for operative rib fixation presented in the literature is age. Outcomes for older patients after severe thoracic trauma have been documented in the literature as poorer than the younger population [37–41, 54]. As a result, one study by de Jong et al. included not only patients with flail chest and/or more than four rib fractures but also those 45 years of age or greater when determining indication for operative rib fixation [6]. Of note, there is no evidence to support using age as an indication for operative rib fixation, and as a result, this is still a theoretical indication.

Contraindications for Rib Fixation and Special Considerations

Although there are no absolute contraindications to operative rib fixation, there are several relative contraindications and special circumstances which should be considered prior to proceeding to surgery.

Location of Fractures

During preoperative planning, it is important to evaluate the location of the rib fractures and the technical feasibility of repair. Additionally, several trials recommend surgically stabilizing only ribs 3–10, given ribs 1–2 and 11–12 are not considered to contribute a significant amount to chest wall stability or pulmonary mechanics [12, 14].

Pulmonary Contusion

The presence or absence of a pulmonary contusion has been found to be an important factor in the decision making regarding whether to offer operative rib fixation. Several studies have shown that patients with pulmonary contusion do not demonstrate the same benefits of operative fixation as those without pulmonary contusion [17, 55–57]. Although the exact physiologic mechanism for decreased efficacy of operative rib fixation in patients with underlying pulmonary contusion is unknown, many authors consider the presence of pulmonary contusion a relative contraindication to surgery. Voggenreiter et al. divided patients into four groups based on the presence or absence of a flail chest and the presence and absence of a pulmonary contusion. He concluded that patients with pulmonary contusions did not benefit from operative rib fixation [56]. That being said, other studies have investigated operative rib fixation in patients with underlying pulmonary contusions and noted some benefit to the procedure including shortened duration of mechanical ventilation. These findings suggest the presence of pulmonary contusions may not play as important a role in operative decision making as previously thought [12, 58, 59]. More research is needed in order to clarify the legitimacy of this relative contraindication.

Traumatic Brain Injury

Another relative contraindication and common exclusion criteria for operative rib fixation is traumatic brain injury. It is unknown whether these patients may benefit from operative rib fixation given the severity of their brain injury will have substantial impact on their overall mortality, ability to wean from mechanical ventilation, need for tracheostomy, and hospital length of stay. This makes interpreting outcomes in this subset of patients extremely difficult. In addition, these patients may not tolerate a long surgical procedure lying flat given concerns for increases in intracranial pressures [12, 14, 15, 45, 59–61].

Timing of Fixation

The final consideration to take into account when determining if operative rib fixation is indicated is the timing of surgery. Although there is no consensus regarding the optimal timing of operative rib fixation, several studies (including the RCTs conducted by Granetzny et al. and Marasco et al.) have suggested operative rib fixation within 72 h of injury offers the most benefit [4, 5, 21, 28]. Although one of the landmark RCTs in this field (by Tanaka et al.) was conducted on patients unable to wean from mechanical ventilation after 5 days [12], many authors believe waiting to proceed to operative fixation until after the patient requires prolonged mechanical ventilation does the patient a disservice since it limits the potential benefit in terms of days on mechanical ventilation and ICU LOS. Furthermore, it can be difficult to predict which patients will require prolonged ventilation or tracheostomy during the first few days post-trauma.

Conclusion

Unfortunately, there are no universal or absolute indications or contraindication for operative rib fixation in the thoracic trauma population. The current recommendation is based on outcomes from three small RCTs, retrospective studies, and expert opinion. The most common and widely accepted indication for operative rib fixation is the presence of a flail chest segment in a patient on mechanical ventilation with no underlying pulmonary contusion. Other relative indications discussed in the literature include severe chest wall deformity, symptomatic rib fracture nonunion, acute pulmonary herniation, failure to wean from mechanical ventilation or poor pulmonary mechanics in a patient with severe chest wall trauma, acute or chronic pain/symptom control, and thoracotomy for another reason. Surgeons should take into consideration possible contraindications for operative repair including the location of rib fractures and presence of either pulmonary contusion or traumatic brain injury.

References

1. Dehghan N, de Mestral C, McKee MD, Schemitsch EH, Nathens A. Flail chest injuries: a review of outcomes and treatment practices from the National Trauma Data Bank. J Trauma Acute Care Surg. 2014;76(2):462–8.
2. Lafferty PM, Anavian J, Will RE, Cole PA. Operative treatment of chest wall injuries: indications, technique, and outcomes. J Bone Joint Surg Am. 2011;93(1):97–110.
3. Swart E, Laratta J, Slobogean G, Mehta S. Operative treatment of rib fractures in flail chest injuries: a meta-analysis and cost-effectiveness analysis. J Orthop Trauma. 2017;31(2):64–70.
4. Nirula R, Mayberry JC. Rib fracture fixation: controversies and technical challenges. Am Surg. 2010;76(8):793–802.
5. Kasotakis G, Hasenboehler EA, Streib EW, et al. Operative fixation of rib fractures after blunt trauma: a practice management guideline from the eastern Association for the Surgery of trauma. J Trauma Acute Care Surg. 2017;82(3):618–26.
6. de Jong MB, Kokke MC, Hietbrink F, Leenen LP. Surgical Management of rib Fractures: strategies and literature review. Scand J Surg. 2014;103(2):120–5.
7. Mayberry JC, Ham LB, Schipper PH, Ellis TJ, Mullins RJ. Surveyed opinion of American trauma, orthopedic, and thoracic surgeons on rib and sternal fracture repair. J Trauma. 2009;66(3):875–9.
8. Kerr-Valentic MA, Arthur M, Mullins RJ, Pearson TE, Mayberry JC. Rib fracture pain and disability: can we do better? J Trauma. 2003;54(6):1058–63. discussion 1063-1054
9. Ziegler DW, Agarwal NN. The morbidity and mortality of rib fractures. J Trauma. 1994;37(6):975–9.

10. Pate JW. Chest wall injuries. Surg Clin North Am. 1989;69(1):59–70.
11. Bottlang M, Long WB, Phelan D, Fielder D, Madey SM. Surgical stabilization of flail chest injuries with MatrixRIB implants: a prospective observational study. Injury. 2013;44(2):232–8.
12. Tanaka H, Yukioka T, Yamaguti Y, et al. Surgical stabilization of internal pneumatic stabilization? A prospective randomized study of management of severe flail chest patients. J Trauma. 2002;52(4):727–32. discussion 732
13. Karev DV. Operative management of the flail chest. Wiad Lek. 1997;50(Suppl 1 Pt 2):205–8.
14. Granetzny A, Abd El-Aal M, Emam E, Shalaby A, Boseila A. Surgical versus conservative treatment of flail chest. Evaluation of the pulmonary status. Interact Cardiovasc Thorac Surg. 2005;4(6):583–7.
15. Marasco SF, Davies AR, Cooper J, et al. Prospective randomized controlled trial of operative rib fixation in traumatic flail chest. J Am Coll Surg. 2013;216(5):924–32.
16. Cataneo AJ, Cataneo DC, de Oliveira FH, Arruda KA, El Dib R, de Oliveira Carvalho PE. Surgical versus nonsurgical interventions for flail chest. Cochrane Database Syst Rev. 2015;7:CD009919.
17. Nirula R, Allen B, Layman R, Falimirski ME, Somberg LB. Rib fracture stabilization in patients sustaining blunt chest injury. Am Surg. 2006;72(4):307–9.
18. Mayberry JC, Kroeker AD, Ham LB, Mullins RJ, Trunkey DD. Long-term morbidity, pain, and disability after repair of severe chest wall injuries. Am Surg. 2009;75(5):389–94.
19. Wu WM, Yang Y, Gao ZL, Zhao TC, He WW. Which is better to multiple rib fractures, surgical treatment or conservative treatment? Int J Clin Exp Med. 2015;8(5):7930–6.
20. Mitchell JD. Blunt chest trauma: is there a place for rib stabilization? J Thorac Dis. 2017;9(Suppl 3):S211–7.
21. Senekjian L, Nirula R. Rib fracture fixation: Indications and Outcomes. *Crit Care Clin.* 2017;33(1):153–65.
22. Coughlin TA, Ng JW, Rollins KE, Forward DP, Ollivere BJ. Management of rib fractures in traumatic flail chest: a meta-analysis of randomised controlled trials. Bone Joint J. 2016;98-B(8):1119–25.
23. Schuurmans J, Goslings JC, Schepers T. Operative management versus non-operative management of rib fractures in flail chest injuries: a systematic review. Eur J Trauma Emerg Surg. 2017;43(2):163–8.
24. Leinicke JA, Elmore L, Freeman BD, Colditz GA. Operative management of rib fractures in the setting of flail chest: a systematic review and meta-analysis. Ann Surg. 2013;258(6):914–21.
25. Mayberry JC, Terhes JT, Ellis TJ, Wanek S, Mullins RJ. Absorbable plates for rib fracture repair: preliminary experience. J Trauma. 2003;55(5):835–9.
26. Campbell N, Conaglen P, Martin K, Antippa P. Surgical stabilization of rib fractures using inion OTPS wraps--techniques and quality of life follow-up. J Trauma. 2009;67(3):596–601.
27. Nirula R, Diaz JJ, Trunkey DD, Mayberry JC. Rib fracture repair: indications, technical issues, and future directions. World J Surg. 2009;33(1):14–22.
28. Fowler TT, Taylor BC, Bellino MJ, Althausen PL. Surgical treatment of flail chest and rib fractures. J Am Acad Orthop Surg. 2014;22(12):751–60.
29. Slater MS, Mayberry JC, Trunkey DD. Operative stabilization of a flail chest six years after injury. Ann Thorac Surg. 2001;72(2):600–1.
30. Morgan-Jones RL, Mackie IG. Non-union of a fracture of the 8th rib. Injury. 1996;27(2):147–8.
31. Ten Duis K, IJpma FF. Surgical treatment of snapping scapula syndrome due to Malunion of rib fractures. Ann Thorac Surg. 2017;103(2):e143–4.
32. Takahara K, Uchiyama S, Nakagawa H, Kamimura M, Ohashi M, Miyasaka T. Snapping scapula syndrome due to malunion of rib fractures: a case report. J Shoulder Elb Surg. 2004;13(1):95–8.
33. Cacchione RN, Richardson JD, Seligson D. Painful nonunion of multiple rib fractures managed by operative stabilization. J Trauma. 2000;48(2):319–21.
34. Richardson JD, Franklin GA, Heffley S, Seligson D. Operative fixation of chest wall fractures: an underused procedure? Am Surg. 2007;73(6):591–6. discussion 596-597

35. François B, Desachy A, Cornu E, Ostyn E, Niquet L, Vignon P. Traumatic pulmonary hernia: surgical versus conservative management. J Trauma. 1998;44(1):217–9.
36. Sirmali M, Türüt H, Topçu S, et al. A comprehensive analysis of traumatic rib fractures: morbidity, mortality and management. Eur J Cardiothorac Surg. 2003;24(1):133–8.
37. Flagel BT, Luchette FA, Reed RL, et al. Half-a-dozen ribs: the breakpoint for mortality. Surgery. 2005;138(4):717–23. discussion 723-715
38. Holcomb JB, McMullin NR, Kozar RA, Lygas MH, Moore FA. Morbidity from rib fractures increases after age 45. J Am Coll Surg. 2003;196(4):549–55.
39. Todd SR, McNally MM, Holcomb JB, et al. A multidisciplinary clinical pathway decreases rib fracture-associated infectious morbidity and mortality in high-risk trauma patients. Am J Surg. 2006;192(6):806–11.
40. Kent R, Woods W, Bostrom O. Fatality risk and the presence of rib fractures. Ann Adv Automot Med. 2008;52:73–82.
41. Shulzhenko NO, Zens TJ, Beems MV, et al. Number of rib fractures thresholds independently predict worse outcomes in older patients with blunt trauma. Surgery. 2017;161(4):1083–9.
42. Chien CY, Chen YH, Han ST, Blaney GN, Huang TS, Chen KF. The number of displaced rib fractures is more predictive for complications in chest trauma patients. Scand J Trauma Resusc Emerg Med. 2017;25(1):19.
43. Majercik S, Wilson E, Gardner S, Granger S, VanBoerum DH, White TW. In-hospital outcomes and costs of surgical stabilization versus nonoperative management of severe rib fractures. J Trauma Acute Care Surg. 2015;79(4):533–538; discussion 538-539.
44. Pieracci FM, Lin Y, Rodil M, et al. A prospective, controlled clinical evaluation of surgical stabilization of severe rib fractures. J Trauma Acute Care Surg. 2016;80(2):187–94.
45. Doben AR, Eriksson EA, Denlinger CE, et al. Surgical rib fixation for flail chest deformity improves liberation from mechanical ventilation. J Crit Care. 2014;29(1):139–43.
46. Karmakar MK, Ho AM. Acute pain management of patients with multiple fractured ribs. J Trauma. 2003;54(3):615–25.
47. Fabricant L, Ham B, Mullins R, Mayberry J. Prolonged pain and disability are common after rib fractures. Am J Surg. 2013;205(5):511–5. discussion 515-516
48. de Moya M, Bramos T, Agarwal S, et al. Pain as an indication for rib fixation: a bi-institutional pilot study. J Trauma. 2011;71(6):1750–4.
49. Khandelwal G, Mathur RK, Shukla S, Maheshwari A. A prospective single center study to assess the impact of surgical stabilization in patients with rib fracture. Int J Surg. 2011;9(6):478–81.
50. Landercasper J, Cogbill TH, Lindesmith LA. Long-term disability after flail chest injury. J Trauma. 1984;24(5):410–4.
51. Beal SL, Oreskovich MR. Long-term disability associated with flail chest injury. Am J Surg. 1985;150(3):324–6.
52. Bhatnagar A, Mayberry J, Nirula R. Rib fracture fixation for flail chest: what is the benefit? J Am Coll Surg. 2012;215(2):201–5.
53. Mouton WLD, Furrer M, Regli B, Ris HB. Long-term follow-up of patients with operative stabilization of a flail chest. Thorac Cardiovasc Surg. 1997;45:242–4.
54. Jones KM, Reed RL, Luchette FA. The ribs or not the ribs: which influences mortality? Am J Surg. 2011;202(5):598–604.
55. Ahmed Z, Mohyuddin Z. Management of flail chest injury: internal fixation versus endotracheal intubation and ventilation. J Thorac Cardiovasc Surg. 1995;110(6):1676–80.
56. Voggenreiter G, Neudeck F, Aufmkolk M, Obertacke U, Schmit-Neuerburg KP. Operative chest wall stabilization in flail chest--outcomes of patients with or without pulmonary contusion. J Am Coll Surg. 1998;187(2):130–8.
57. Kishikawa M, Yoshioka T, Shimazu T, Sugimoto H, Sugimoto T. Pulmonary contusion causes long-term respiratory dysfunction with decreased functional residual capacity. J Trauma. 1991;31(9):1203–1208; discussion 1208-1210.
58. Zhang Y, Tang X, Xie H, Wang RL. Comparison of surgical fixation and nonsurgical management of flail chest and pulmonary contusion. Am J Emerg Med. 2015;33(7):937–40.

59. Althausen PL, Shannon S, Watts C, et al. Early surgical stabilization of flail chest with locked plate fixation. J Orthop Trauma. 2011;25(11):641–7.
60. Solberg BD, Moon CN, Nissim AA, Wilson MT, Margulies DR. Treatment of chest wall implosion injuries without thoracotomy: technique and clinical outcomes. J Trauma. 2009;67(1): 8–13. discussion 13
61. Xu JQ, Qiu PL, Yu RG, Gong SR, Ye Y, Shang XL. Better short-term efficacy of treating severe flail chest with internal fixation surgery compared with conservative treatments. Eur J Med Res. 2015;20:55.

9 Slipping Rib Syndrome and Other Causes of Chest Wall Pain

Marisa Gasparri and Mario Gasparri

Chest pain is one of the most common reasons for seeking medical attention, affecting 20–40% of the general population during their lifetime [1]. In 2013 there were six million visits to the emergency room for evaluation of nontraumatic chest pain [2], and 1.5% of all primary care office visits are for the evaluation of chest pain [3, 4]. While significant cardiovascular disease is usually the greatest concern, depending on the emergency room or ambulatory clinic setting, up to 80% of these patients have a "benign" cause for their chest pain. In those with a "benign" cause, half will be classified as having "chest wall syndrome," a broad term encompassing all causes of musculoskeletal pain [3, 5]. Many of these patients, once a life-threatening disorder has been excluded, remain undiagnosed yet continue to experience symptoms [6]. The purpose of this chapter is to review some of the bone- and cartilage-related causes of chest wall syndrome focusing on the slipping rib syndrome, an easily curable and often overlooked cause of recurrent chest pain, as well as briefly reviewing some of the other pathologies the practitioner may encounter.

Anatomy

The thoracic cage is formed by the 12 thoracic vertebrae, the 12 pairs of ribs and costal cartilages, and the sternum anteriorly. All 12 ribs articulate posteriorly with the vertebrae. The "true" ribs, 1–7, articulate directly with the sternum anteriorly via their costal cartilages, while the "floating" ribs, 11–12, have no anterior

M. Gasparri
University of Wisconsin–Madison, Madison, WI, USA
e-mail: mgasparri@wisc.edu

M. Gasparri (✉)
Division of Thoracic Surgery, Medical College of Wisconsin, Milwaukee, WI, USA

M. de Moya, J. Mayberry (eds.), *Rib Fracture Management*,
https://doi.org/10.1007/978-3-319-91644-6_9

attachment to the sternum or to each other. The "false" ribs, 8–10, are attached to the sternum through the costal cartilages of another rib. These attachments, known as interchondral joints, are made up of a fibrous mesh surrounding a synovial membrane and its joint [7, 8]. They allow for the desired mobility during respiration but are sturdy enough to maintain alignment and spacing of each of these ribs.

At the inferior aspect of the sternum is the xiphoid process (Fig. 9.1). The xiphoid process is the most variable chest wall structure and can assume a variety of shapes including broad or narrow, thick or thin, pointed or bifid, and curved or straight. It begins as a cartilaginous structure which becomes ossified in the adult, and it is attached to the sternum by fibrocartilage. The inferior aspect of the seventh rib articulates with the superior lateral margin of the xiphoid.

Associated with each rib is the intercostal bundle consisting of the vein, artery, and nerve (Fig. 9.2). The intercostal bundles travel in the intercostal spaces along the underside of the inferior aspect of each rib. The nerve is the most inferior structure in the bundle and, for much of its course, travels within the layers of the intercostal muscles. As the nerves continue to travel from posterior to anterior and the innermost intercostal muscle layer thins, the lower intercostal nerves (7–11) are

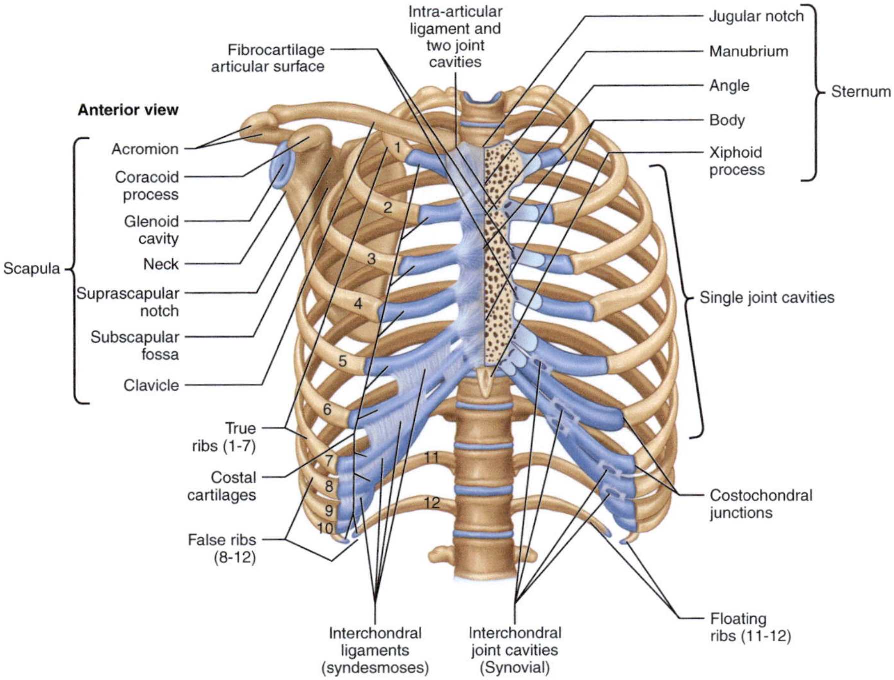

Fig. 9.1 Bony anatomy of the chest wall

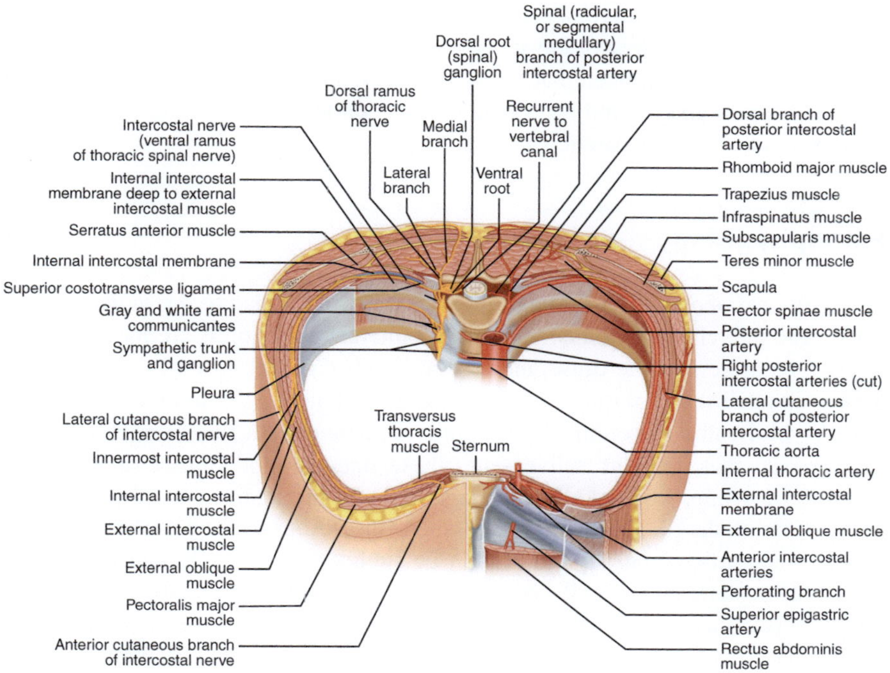

Fig. 9.2 Anatomy of the intercostal nerve

then found between the parietal pleura and posterior aspect of the intercostal muscle prior to then crossing the posterior surfaces of the costal cartilages and continuing on to the abdominal wall.

Slipping Rib

History

Slipping rib syndrome, also known as Cyriax syndrome, rib-tip syndrome, and painful rib syndrome, among others [9–11], was first described in 1919 by Cyriax [9]. In 1922 Davies-Colley coined the term "slipping rib" and reported on two patients with incapacitating subcostal pain, one of whom had undergone a negative laparotomy in efforts to relieve symptoms. He felt this was due to "undue mobility" of the costal cartilages, and after resection, both patients were completely asymptomatic [12]. Since then hundreds of reports have appeared in the literature, yet slipping rib syndrome remains a poorly understood and underappreciated diagnosis and one which "seems to be regarded with skepticism by some clinicians who happily accept other forms of nerve compression e.g. carpal tunnel syndrome" [13]. In fact, following a published report describing six cases and reviewing the literature to dates, the editorial comment cautioned that "...perhaps psychiatric evaluation should also be part of the workup...." And "...this diagnosis, while probably valid on occasion, must be made with due skepticism" [14].

Pathophysiology

As mentioned above, the costal cartilages of ribs 8–10 attach to the more superior rib via the interchondral joints. When these interchondral joints are interrupted or injured, regardless of the cause, the cartilage ends can curl upward and inward to either rub against or get trapped behind the more superior rib. In an elegant study done by McBeath, he examined the eighth, ninth, and tenth ribs from 20 cadavers and showed that, in their native state, these interchondral fibrous attachments, although very thin, "did not allow the eighth, ninth, and tenth rib cartilage to come in contact with the rib above"; however when "the fibrous tissue between them was incised, the cartilaginous tip could then be subluxated inferiorly and posteriorly until it lay posteriorly to the rib above." He also showed that at the rib tips, the intercostal muscle mass was very thin [10]. Based on this study as well as others [15], the mechanism of this syndrome is an injury of these interchondral joints allowing upward motion of the cartilage and rib. As these mobile cartilages move upward, the intercostal nerve, which is particularly vulnerable in this location due to its location and the decreased intercostal muscle mass, is compressed and irritated, resulting in pain. As further support that the pathophysiology of this syndrome is nerve compression and irritation rather than any underlying abnormality or inflammation of the interchondral joints themselves, Holmes showed that pathologic examination of the resected specimens displayed no particular histologic abnormality [15].

Prevalence

The prevalence of this disorder is unknown. Scott reported on 76 consecutive patients diagnosed with slipping rib syndrome in a general medical clinic which accounted for 3% of all new patient visits, while Wright found that 5% of new patients referred to a gastroenterology clinic were found to have this syndrome [11, 16]. These numbers are obviously skewed by the referral patterns and patient population seen in these clinics, and while they may give some idea on how common this disorder can be in an everyday practice, there are no population-based studies from which to draw any meaningful conclusions.

Similarly, true age and gender distribution are not known. While some authors have stated that it is more common in middle age [14, 17], this is probably more reflective of their particular patient population as a review of the literature shows that this syndrome has been reported relatively equally across all age ranges from as young as age 7 to as old as age 86 [18, 19]. Likewise, this syndrome appears to be equally distributed between genders.

Symptoms

What has been consistent among all reports is patient presentation. All patients present with pain which is typically unilateral. The pain is located in the subcostal region and radiates anteriorly. It is usually felt as somatic pain although in some reports it has been perceived as visceral pain which has been confused with biliary colic or some other intraabdominal process. The pain is described as a sharp and stabbing pain, and while the intensity of the pain can vary, in the majority of patients, it is severe and can be debilitating. The author has had several patients who have suffered syncopal episodes due to the sudden onset of acute severe pain. The pain is exacerbated by certain movements, most commonly twisting and bending. The patients become much attuned to the instigating factor and are very careful to avoid these maneuvers. Often the patient will feel a "click or pop" in the lower rib cage with these episodes. Relief is most often achieved by lying supine, often with a pillow under the back, and taking deep breaths. The savvy patient can actually lie still and reduce the slipped rib back into position to relieve their pain.

Diagnosis

Diagnosis is made by the very classic history and confirmed by physical exam. As outlined above, these patients describe a very consistent pain complex and are very detailed with regard to which maneuvers elicit their pain. Many patients can relate a "direct" traumatic event which predated their symptoms such as a fall, auto accident, or other chest wall impact. Other more subtle "indirect" traumatic causes which have been consistently described include sudden flexion, extension or twisting of the body, repetitive strenuous motions (swimmers, gymnasts, golfers), or forced compression and/or expansion of the chest (childbirth, forceful coughing).

On physical examination of the chest wall, the patient's pain will be elicited by simple palpation and gentle pressure of the affected rib and costal cartilage. The unaffected side serves as a control. If the patient can tolerate it, the "click or pop" can also be elicited with more direct pressure, and the author has had several patients who will manipulate the rib themselves in an effort to display their abnormality to the examiner. The classic diagnostic maneuver is the "hooking maneuver" described by Heinz and Zavala [20] in which the examiner hooks his or her fingers behind the lower costal margin and pulls anteriorly and superiorly, reproducing the patient's pain. In the author's opinion, this test is often unnecessary as the diagnosis can often be made short of this maneuver, and this maneuver can elicit pain even in the unaffected patient. Note that the physical exam findings should be reliable and reproducible and palpation over the exact same area should always elicit the exact same pain. If the exam changes or the findings be more vague, then the diagnosis should be questioned.

The diagnosis is typically made on history and physical exam, and imaging adds very little. If the diagnosis is in doubt, however, ultrasound has been described and can show the subluxing cartilage and rib [21, 22]. In general, the role of imaging is mainly to rule out other potential etiologies in the differential diagnosis of the patient's symptoms.

Treatment

Various treatment modalities have been described, from least invasive to most invasive, only one of which truly addresses the underlying anatomic abnormality.

The least invasive form of treatment with some improved outcomes is simple recognition of the problem combined with reassurance and counseling. In patients with minimal symptoms, which one could argue are patients less likely to actually seek out and/or need medical attention, this may prove useful. In patients with more moderate to severe symptoms, while success has been reported, a careful review of these reports shows that the patients merely alter their lifestyle and avoid the movements and positions which provoke their pain [11, 19]. In the only study which gives some idea of the natural history of the disease, 76 patients with the diagnosis of slipping rib syndrome who were treated expectantly were subsequently surveyed. Fifty-six percentage responded, and 70% of these patients still had pain after a mean of 8 years [11].

Others have described injections of the interspace, most often using local anesthesia [17, 23] although steroids [24] and botulinum toxin [25] have also been described. These modalities have met with inconsistent results and poor follow-up.

The only truly curative modality reported in the literature is surgical resection of the affected cartilage. Hundreds of cases have been documented [8, 10, 12–15, 18, 26–29], and all report complete relief of symptoms following excision with many of these having follow-up measured in years.

The surgical approach consists of careful preoperative marking in the awake patient over the point of maximal tenderness. The eighth, ninth, or tenth cartilage can be affected. The patient is then positioned supine, often with a bump under the affected side to maximize exposure of the anterolateral chest wall. The author prefers general anesthesia although this could be done under monitored anesthesia care in the carefully selected patient. An incision is made overlying the affected cartilage and the covering musculature divided in line with its fibers. Often, just by doing this, the affected cartilage will “pop up” into the field. The anterior tip is then grasped, and the cartilage is noted to be freely mobile and can often be stood almost straight up to facilitate the dissection. Dissection continues posteriorly to the costochondral junction, and then the cartilage with or without a small portion of the rib is resected. Dissection can usually be completed extrapleurally; however if the pleura is inadvertently entered, the residual air can be evacuated just prior to closure. The wound is then closed in layers and no drain is necessary. The vast majority of these can be done as outpatient surgery although there will be the occasional patient requiring admission overnight for pain control.

Costochondritis

Costochondritis is a relatively common condition, and as its name implies, it is an inflammation of the costochondral joints. This condition can affect children as well as adults, and males and females are equally affected. While exact incidence is unknown, reports have shown that this is the diagnosis in up to one-third of patients undergoing evaluation for chest pain depending on the setting [5, 30].

The etiology is unclear, but often the patient will describe an antecedent illness associated with strenuous coughing or recent strenuous exercise or lifting. The presenting complaint is always pain although it can vary from dull and heavy to sharp and stabbing. The pain tends to be limited to the chest wall without radiation and is often exacerbated by pushing, pulling, or coughing. The pain involves several cartilages, usually the second through fifth, and is most often unilateral with a tendency toward the left side.

On physical exam, the patients have pain on palpation of the affected cartilages. There is no swelling, warmth, or erythema, and there is no instability of the chest wall. Certain physical exam maneuvers such the "crowing rooster" in which the examiner asks the patient to clasp their hands behind their head and, while standing behind the patient, exerts superior and posterior pressure at the elbows can reproduce the pain, but they are by no means pathognomonic of this disorder [31].

Diagnosis is based on the history and exam findings, and there are no diagnostic studies specific for costochondritis. Some reports have shown elevated erythrocyte sedimentation rates or abnormal nuclear scans associated with costochondritis; however these findings are inconsistent and nonspecific, and diagnostic studies are most helpful in ruling out other potential etiologies [30, 32–35]. Along these lines, since most patients present with left-sided chest pain exacerbated by activity, ruling out an underlying cardiac cause is the most common workup these patients undergo. One important note, however, is that while chest wall tenderness to palpation is classic with costochondritis and unusual with an acute coronary syndrome, both can coexist, and one emergency room study showed that 6% of patients diagnosed with costochondritis also had an acute myocardial infarct [30].

Treatment is supportive with reassurance and mild analgesics (Tylenol or nonsteroidal anti-inflammatory agents) combined with stretching exercises [31], and symptoms typically resolve within 12 months with recurrences uncommon. One interesting report suggested that early rheumatologic intervention and use of steroid injection and/or sulfasalazine may accelerate recovery and avoid hospital admission although this was a small group of patients and has not been reproduced [36].

Tietze's Syndrome

Tietze's syndrome is an inflammatory condition involving the costal cartilage. It is similar in some ways to costochondritis (see above), and many mistakenly use the terms interchangeably when, in fact, Tietze's syndrome and costochondritis are two distinct entities. Tietze's syndrome is distinguished from costochondritis in that the pain is associated with swelling of the area and 80% of the time [37–39] it affects only a single costal cartilage. Its prevalence is not known, but it is much less common than costochondritis. It affects all age groups with no predilection for gender [37, 38]. The second or third costal cartilage is involved most often with right and left side affected equally.

When first described by Tietze in 1921 postwar Germany [40], he felt it was due to war-time malnutrition or possibly tuberculosis; however large reviews of this syndrome have disproven both of these theories [37, 39, 41]. Other causes have since been postulated including viral illness [42], trauma to the cartilage [43, 44], surrounding ligamentous injury [45], or respiratory infection [41]; however none of these theories have been consistently identified, and the etiology remains obscure. Surgical exploration has similarly not offered much insight into the etiology as gross surgical findings have been unrevealing, and pathologic examination of the affected cartilage and even the surrounding soft tissue have shown no specific histologic abnormality [37, 39].

Patients present with a painful swelling of the involved costal cartilage. The pain typically precedes the swelling, and its onset can be sudden or gradual, and its intensity can be severe or mild. The pain is often exacerbated by movement or coughing, and its mode of presentation often dictates the setting of evaluation whether it be the emergency with concerns of angina or in the oncologist or surgeon's office with concerns of tumor. The swelling is spindle shaped and confined to the affected cartilage and is not associated with warmth or erythema.

Diagnosis is made by the typical history and physical exam and after excluding other causes. Routine investigations do not show abnormalities, and there are no tests which are diagnostic for Tietze's syndrome. There have been reports of ultrasonography [46, 47] or MRI [48] being useful although further experience with these techniques will be needed to determine their reliability. In general, laboratory and imaging studies are used to rule out other potential etiologies whether they be cardiac, infectious, or neoplastic, and the diagnosis of Tietze's syndrome is made when other possibilities have been excluded.

This syndrome is typically self-limited with the swelling and pain reaching a maximum and then subsiding, typically over the course of a few weeks to a few months. Although there has been a report of a case lasting 8 years [49], this is extremely unusual, and failure to show improvement in the expected timeframe, or more ominous, continued growth over this period, warrants further workup and/or biopsy in order to exclude an underlying malignant process [50, 51].

As the symptoms are typically self-limited, traditional treatment consists of reassurance, analgesics (Tylenol, nonsteroidal anti-inflammatory drugs), and heat and occasional steroid injection if the symptoms are more severe [47, 52, 53]. One recent article suggests prolotherapy may provide better pain relief and accelerate recovery [54].

Xiphodynia

Xiphodynia or the painful xiphoid syndrome refers to pain in the region of the xiphoid process. Although first described in 1712 [55], it has received little attention in the literature, and the majority of publications consist of case reports or small series. The incidence and prevalence of this disorder are unknown, and while many authors consider it an uncommon problem [56–58], Lipkin noted a prevalence of 2% in a general hospital ward population and felt it is "far more common than is generally appreciated" [55].

The etiology is also unclear, and several mechanisms have been postulated including inflammation due to repetitive mechanical stress [56, 59]; abnormal angulation of the xiphoid, congenital or acquired, which makes it more susceptible to injury [57, 60]; trauma, blunt or surgical [61, 62]; or "Tietze's syndrome" of the xiphisternal joint [32, 63]. None of these theories has been consistently identified in all patients.

Most patients present with epigastric pain that is subacute or chronic, usually having gotten progressively worse over the course of weeks or months. Very rarely do patients present acutely. The pain is typically a dull, achy pain which tends to be limited to the epigastric area with occasional radiation to the substernal or abdominal area. When radiating to these areas the patient often experiences nausea. The pain is exacerbated by direct pressure over the area such as when lying prone or hugging someone or by traction on the xiphoid such as when twisting, bending, or deep breathing. On exam, light palpation over the xiphoid will reproduce their pain as well as the radiation and associated nausea.

The diagnosis is based on history and physical exam, and there are no tests pathognomonic for this disorder. It is most often overlooked as the practitioner focuses on cardiac or abdominal reasons for their complaints and patients have undergone cardiac or abdominal procedures without relief of symptoms before the diagnosis is made [55, 56, 64].

Once the diagnosis is made, most authors advise reassurance and mild analgesics. In the cases of repetitive mechanical stress, cessation of these activities has led to resolution of symptoms [56, 57]. If pain is more severe, injections with local anesthesia [55, 57, 63] or corticosteroids [63] and ultrasound therapy [65] have been described. When pain persists or if anatomic abnormalities are present, surgical excision has been reported, and all reports have documented complete resolution of symptoms [57, 59, 60, 64].

References

1. Ruigómez A, Rodríguez LA, Wallander MA, Johansson S, Jones R. Chest pain in general practice: incidence, comorbidity and mortality. Fam Pract. 2006;23(2):167–74. https://doi.org/10.1093/fampra/cmi124.
2. National Hospital Ambulatory Medical Care Survey: 2009 emergency dept summary tables–table 10. 2009. Available at: https://www.cdc.gov/nchs/data/ahcd/nhamcs_emergency/2013_ed_web_tables.pdf. Accessed 25 Mar 2017.
3. Bösner S, Becker A, Hani MA, Keller H, Sönnichsen AC, Karatolios K, Schaefer JR, Haasenritter J, Baum E, Donner-Banzhoff N. Chest wall syndrome in primary care patients with chest pain: presentation, associated features and diagnosis. Fam Pract. 2010;27(4):363–9. https://doi.org/10.1093/fampra/cmq024.
4. Nilsson S, Scheike M, Engblom D, Karlsson LG, Mölstad S, Åkerlind I, Örtoft K, Nylander E. Chest pain and ischaemic heart disease in primary care. Br J Gen Pract. 2003;53(5):378–82.

5. Wise CM. Chest wall syndromes. Curr Opin Rheumatol. 1994;6(2):197–202.
6. Stochkendahl MJ, Christensen HW. Chest pain in focal musculoskeletal disorders. Med Clin North Am. 2010;94(2):259–73. https://doi.org/10.1016/j.mcna.2010.01.007.
7. Telford KM. The slipping rib syndrome. Can Med Assoc J. 1950;62(5):463–5.
8. Bass J, Pan HC, Fegelman RH. Slipping rib syndrome. J Natl Med Assoc. 1979;71(9):863–5.
9. Cyriax EF. On various conditions that may simulate the referred pains of visceral disease, and a consideration of these from the point of view of cause and effect. Practitioner. 1919;102:314–22.
10. McBeath AA, Keene JS. The rib-tip syndrome. J Bone Joint Surg Am. 1975;57(6):795–7.
11. Scott EM, Scott BB. Painful rib syndrome—a review of 76 cases. Gut. 1993;34(7):1006–8.
12. Davies-Colley R. Slipping rib. Br Med J. 1922;1(3194):432.
13. Copeland GP, Machin DG, Shennan JM. Surgical treatment of the 'slipping rib syndrome'. Br J Surg. 1984;71(7):522–3.
14. Spence EK, Rosato EF. The slipping rib syndrome. Arch Surg. 1983;118(11):1330–2.
15. Holmes JF. A study of the slipping rib cartilage syndrome. N Engl J Med. 1941;224:928–32.
16. Wright JT. Slipping-rib syndrome. Lancet. 1980;2(8195 pt 1):632–4.
17. Kingsley RA. A little-known cause of chest pain in a 14-year-old athlete. J Pediatr Health Care. 2014;28(6):555–8. https://doi.org/10.1016/j.pedhc.2014.04.005.
18. Saltzman DA, Schmitz ML, Smith SD, Wagner CW, Jackson RJ, Harp S. The slipping rib syndrome in children. Paediatr Anaesth. 2001;11(6):740–3.
19. Arroyo JF, Vine R, Reynaud C, Michel JP. Slipping rib syndrome: don't be fooled. Geriatrics. 1995;50(3):46–9.
20. Heinz GJ, Zavala DC. Slipping rib syndrome. JAMA. 1977;237(8):794–5.
21. Meuwly JY, Wicky S, Schnyder P, Lepori D. Slipping rib syndrome: a place for sonography in the diagnosis of a frequently overlooked cause of abdominal or low thoracic pain. J Ultrasound Med. 2002;21(3):339–43.
22. Zbojniewicz AM. US for diagnosis of musculoskeletal conditions in the young athlete: emphasis on dynamic assessment. Radiographics. 2014;34(5):1145–62. https://doi.org/10.1148/rg.345130151.
23. Waldman SD. Slipping rib syndrome. In: Waldman SD, editor. Atlas of uncommon pain syndromes. 3rd ed. Philadelphia, PA: Saunders; 2013. p. 198–9.
24. Robb LG, Robb MP, Robb PM. The slipping rib syndrome: an overlooked cause of abdominal pain. Pract Pain Manag. 2013;13(7):1–3.
25. Pirali C, Santus G, Faletti S, De Grandis D. Botulinum toxin treatment for slipping rib syndrome: a case report. Clin J Pain. 2013;29(10):e1–3. https://doi.org/10.1097/AJP.0b013e318278d497.
26. Porter GE. Slipping rib syndrome: an infrequently recognized entity in children: a report of three cases and review of the literature. Pediatrics. 1985;76(5):810–3.
27. Fu R, Iqbal CW, Jaroszewski DE, St Peter SD. Costal cartilage excision for the treatment of pediatric slipping rib syndrome. J Pediatr Surg. 2012;47(10):1825–7. https://doi.org/10.1016/j.jpedsurg.2012.06.003.
28. Gould JL, Rentea RM, Poola AS, Aguayo P, St Peter SD. The effectiveness of costal cartilage excision in children for slipping rib syndrome. J Pediatr Surg. 2016;51(12):2030–2. https://doi.org/10.1016/j.jpedsurg.2016.09.032.
29. van Delft EA, van Pul KM, Bloemers FW. The slipping rib syndrome: a case report. Int J Surg Case Rep. 2016;23:23–4. https://doi.org/10.1016/j.ijscr.2016.04.009.
30. Disla E, Rhim HR, Reddy A, Karten I, Taranta A. Costochondritis. A prospective analysis in an emergency department setting. Arch Intern Med. 1994;154(21):2466–9.
31. Ayloo A, Cvengros T, Marella S. Evaluation and treatment of musculoskeletal chest pain. Prim Care. 2013;40(4):863–87. https://doi.org/10.1016/j.pop.2013.08.007.
32. Fam AG, Smythe HA. Musculoskeletal chest wall pain. CMAJ. 1985;133(5):379–89.
33. Mendelson G, Mendelson H, Horowitz SF, Goldfarb CR, Zumoff B. Can (99m)technetium methylene diphosphonate bone scans objectively document costochondritis? Chest. 1997;111(6):1600–2.

34. Ikehira H, Kinjo M, Nagase Y, Aoki T, Ito H. Acute pan-costochondritis demonstrated by gallium scintigraphy. Br J Radiol. 1999;72(854):210–1. https://doi.org/10.1259/bjr.72.854.10365077.
35. Miller JH. Accumulation of gallium-67 in costochondritis. Clin Nucl Med. 1980;5(8):362–3.
36. Freeston J, Karim Z, Lindsay K, Gough A. Can early diagnosis and management of costochondritis reduce acute chest pain admissions? J Rheumatol. 2004;31(11):2269–71.
37. Kayser HL. Tietze's syndrome; a review of the literature. Am J Med. 1956;21(6):982–9.
38. Levey GS, Calabro JJ. Tietze's syndrome: report of two cases and review of the literature. Arthritis Rheum. 1962;5(3):261–9.
39. Landon J, Malpas JS. Tietze's syndrome. Ann Rheum Dis. 1959;18:249–54.
40. Tietze A. Ueber eine eigenartige Haufung von Fallen mit Dystrophie der Rippenknorpel. Berl Klin Wschr. 1921;58:829–31.
41. Geddes AK. Tietze's syndrome. Can Med Assoc J. 1945;53:571–3.
42. Gill GV. Epidemic of Tietze's syndrome. Br Med J. 1977;2(6085):499.
43. Gukelberger M. Tietze's syndrome and its significance in differential diagnosis. Schweiz Med Wochenschr. 1953;83(12):288–9.
44. Coventry MB. Thoracic pain of musculoskeletal origin. Proc Staff Meet Mayo Clin. 1956;31(1):6–9.
45. Motulsky A, Rohn RJ. Tietze's syndrome: cause of chest pain and chest wall swelling. J Am Med Assoc. 1953;152(6):504–6.
46. Martino F, D'Amore M, Angelelli G, Macarini L, Cantatore FP. Echographic study of Tietze's syndrome. Clin Rheumatol. 1991;10(1):2–4.
47. Kamel M, Kotob H. Ultrasonographic assessment of local steroid injection in Tietze's syndrome. Br J Rheumatol. 1997;36(5):547–50.
48. Volterrani L, Mazzei MA, Giordano N, Nuti R, Galeazzi M, Fioravanti A. Magnetic resonance imaging in Tietze's syndrome. Clin Exp Rheumatol. 2008;26(5):848–53.
49. Ausubel H, Cohen BD, Ladue JS. Tietze's disease of eight years' duration. N Engl J Med. 1959;261(4):190. https://doi.org/10.1056/NEJM195907232610408.
50. Fioravanti A, Tofi C, Volterrani L, Marcolongo R. Malignant lymphoma presenting as Tietze's syndrome. Arthritis Rheum. 2002;47(3):229–30. https://doi.org/10.1002/art.10401.
51. Thongngarm T, Lemos LB, Lawhon N, Harisdangkul V. Malignant tumor with chest wall pain mimicking Tietze's syndrome. Clin Rheumatol. 2001;20(4):276–8.
52. Härkönen M. Tietze's syndrome. Br Med J. 1977;2(6094):1087–8.
53. Semble EL, Wise CM. Chest pain: a rheumatologist's perspective. South Med J. 1988;81(1):64–8.
54. Şentürk E, Şahin E, Serter S. Prolotherapy: an effective therapy for Tietze syndrome. J Back Musculoskelet Rehabil. 2017. https://doi.org/10.3233/BMR-159269.
55. Lipkin M, Fulton LA, Wolfson EA. The syndrome of the hypersensitive xiphoid. N Engl J Med. 1955;253(14):591–7. https://doi.org/10.1056/NEJM195510062531403.
56. Ugurlar OY, Ugurlar M, Ozel A, Erturk SM. Xiphoid syndrome: an uncommon occupational disorder. Occup Med. 2014;64:64–6.
57. Maigne JY, Vareli M, Rousset P, Cornelis P. Xiphodynia and prominence of the xyphoid process. Value of xiphosternal angle measurement: three case reports. Joint Bone Spine. 2010;77(5):474–6. https://doi.org/10.1016/j.jbspin.2010.04.009.
58. Howell J. Xiphoidynia: an uncommon cause of exertional chest pain. Am J Emerg Med. 1990;8(2):176.
59. Sano A, Inui M. Xiphoidectomy for xiphoid process-induced pain in a surfer. Asian Cardiovasc Thorac Ann. 2015;23(9):1116–8. https://doi.org/10.1177/0218492315589198.
60. Hogerzeil DP, Hartholt KA, de Vries MR. Xiphoidectomy: a surgical intervention for an underdocumented disorder. Case Rep Surg. 2016;2016:9306262. https://doi.org/10.1155/2016/9306262.
61. Enomoto N, Tayama K, Kohno M, Otsuka H, Yokose S, Kosuga K. Postoperative elongation of the xiphoid process – report of a case. Ann Thorac Cardiovasc Surg. 2011;17(3):307–9.
62. Waldman SD. Xiphodynia. In: Waldman SD, editor. Atlas of uncommon pain syndromes. 3rd ed. Philadelphia, PA: Saunders; 2013. p. 193–4.

63. Jelenko C, Cowan GS. Perichondritis (Tietze's syndrome) at the xiphisternal joint: a mimic of severe disease. JACEP. 1977;6(12):536–42.
64. Tanaka Y, Sakata K, Waseda Y, Fujimura T, Yamada K, Oyama T, Kawashiri MA, Yamagishi M. Xiphodynia mimicking acute coronary syndrome. Intern Med. 2015;54(12):1563–6. https://doi.org/10.2169/internalmedicine.54.3449.
65. Simpson JK, Hawken E. Xiphodynia: a diagnostic conundrum. Chiropr Osteopat. 2007;15:13. https://doi.org/10.1186/1746-1340-15-13.

Rib Fracture Management Algorithm for the Traumatically Injured, Non-intubated Patient

10

Walter Biffl and Frank Zhao

The goal of this algorithm is to recognize patients who are high risk for respiratory compromise and guide analgesia, pulmonary support, and clinical monitoring. Implementation of a structured rib fracture management algorithm has been shown to decrease ICU and hospital lengths of stay, pneumonia, and mortality [1].

1. *Risk factors*—The main morbidities of traumatic rib fractures are the impairment of respiratory function, the subsequent development of pneumonia, and the resulting respiratory failure. Multiple published risk factors can be divided into three main categories for easier evaluation: (a) the severity of the injury, (b) the functional impact of the injury, and (c) the patient's physiologic reserve (Fig. 10.1).
2. *Severity of injury*—One of the first steps in evaluation should be to determine injury severity. Several injury patterns have been previously studied and shown to be risk factors for worse outcomes. Four or more rib fractures have been shown to increase morbidity in patients as young as 45 years of age [2]. Flail chest, defined as three or more consecutive ribs fractured in two or more places, significantly increases both morbidity and mortality [3]. Even in the absence of flail chest, three or more rib fractures with bicortical displacement can result in significant loss of lung volumes leading to respiratory compromise. Finally, the severity of pulmonary contusions by volume correlates with both need for ventilator support and length of time on mechanical ventilation [4]. Significant contusion of greater than 20% of lung parenchyma should be considered a risk factor.

W. Biffl (✉)
Taruma and Acute Care Surgery, Scripps Memorial Hospital, La Jolla, CA, USA

Scripps Memorial Hospital, La Jolla, CA, USA
e-mail: walt@biffl.com

F. Zhao
Acute Care Surgery, The Queen's Medical Center, Honolulu, HI, USA

Department of Surgery, John A. Burns School of Medicine, University of Hawaii, Honolulu, HI, USA
e-mail: fzhao@queens.org

M. de Moya, J. Mayberry (eds.), *Rib Fracture Management*,
https://doi.org/10.1007/978-3-319-91644-6_10

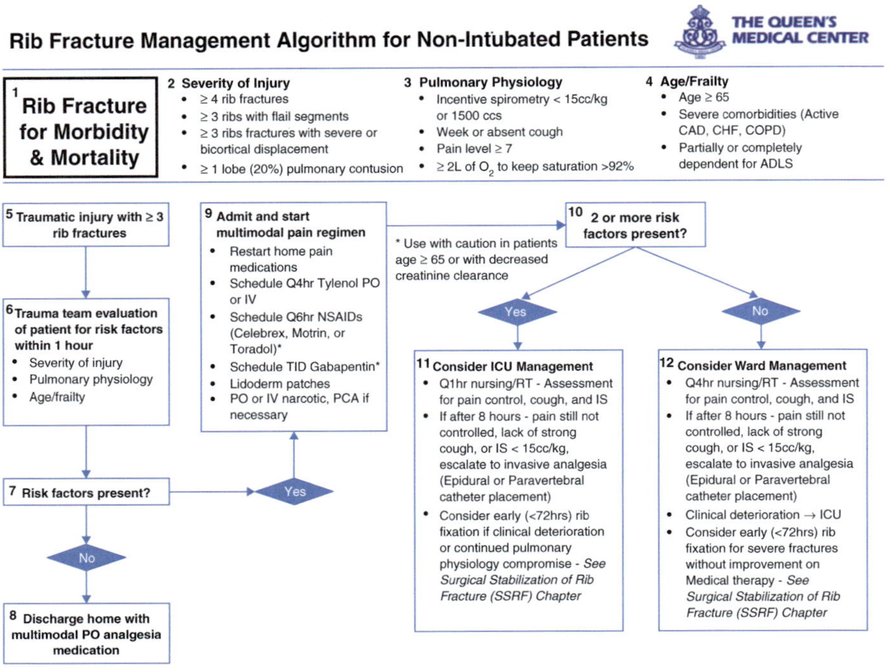

Fig. 10.1 Rib fracture management algorithm for non-intubated patients

3. *Pulmonary physiology*—The second group of risk factors relates to the impact of the injury on the patient's pulmonary physiology. Individual patients differ in their tolerance to pain. Uncontrolled pain from rib fractures leads to respiratory splinting, decreased inspiratory volume, and inadequate alveolar expansion. Similarly, uncontrolled pain leads to a decline in respiratory secretion clearance by a weakened or absent cough. Patients who have pain levels of 7 or higher are considered to be poorly controlled [5]. Finally, the need for supplemental oxygen to maintain normal physiologic oxygen saturation is a sign of pulmonary function compromise. Although multiple thresholds have been suggested, we have chosen to identify the risk factor as a patient with normal baseline saturation with a new requirement of 2 or more liters of supplemental oxygen to maintain saturation at 92% or greater.
4. *Age/frailty*—This third group categorizes risk factors that decrease the patient's physiologic reserve to tolerate an injury. The geriatric population has been shown to have higher rates of morbidity and mortality than younger adults for the same severity of injury [6, 7]. Likewise, patients who have active cardiopulmonary disease have decreased ability to compensate. Finally, although there are many frailty scoring systems, the patient's dependence for activities of daily living is a quick and straightforward way of determining frailty risk [8]. The most recent Western Trauma Association (WTA) guidelines regard frailty and underlying cardiopulmonary disease as additional risk factors rather than age alone [9].

5. Evaluation begins with the identification of a traumatically injured patient with three or more rib fractures [10].
6. The three categories of risk factors should be promptly evaluated within 1 h. This should be taken in context with the full extent of traumatic injuries of the patient.
7. The first major decision point considers whether the patient can be safely discharged from the emergency department. After a full evaluation of possible risk factors, the traumatically injured patient with rib fractures should be considered for discharge home if none are present.
8. Discharge of the patient should be accompanied by education regarding the importance of warning signs of respiratory compromise. Additionally, a multimodal pain regimen should be prescribed to include not only a narcotic if indicated but also simultaneous nonnarcotic adjuncts such as acetaminophen and a nonsteroidal anti-inflammatory drug (NSAID) [11]. A multimodal pain regimen has been shown to decrease reliance on narcotics and possibly offers improved pain control [12].
9. If there are any risk factors present, then the patient should be admitted for inpatient observation and/or treatment. Pain is one of the main factors for decreased pulmonary physiology; therefore, we suggest early aggressive multimodal pain therapy [11]. This should include resumption of any home pain medications that the patient is currently taking. Nonnarcotic adjuncts such as scheduled acetaminophen and NSAIDs should be selected if the patient has no contraindications to these medications. Selective COX-2 inhibitors celecoxib PO and parecoxib IV offer the advantage of analgesia without risk of GI ulceration. Ketorolac has also been shown to decrease morbidities such as pneumonias and shorten length of stay after rib fractures [13, 14]. Although gabapentin and Lidoderm patches have not shown consistent benefit, their risk/benefit ratio warrants consideration for use in rib fractures. Finally, narcotic medications should be routinely given as part of the multimodal regimen. The route (PO, IV, PCA) should be determined based on the severity of injury and the patient's pain levels.
10. The second major decision point is determination of ICU need. Multiple authors have previously used strictly age and number of rib fractures as ICU admission criteria [2, 5, 9]. In patients with three or more rib fractures, while age ≥65 is a definite risk factor, using this as the sole determinate in ICU admission can result in overtriage of noncritical patients to the ICU and undertriage of patients with other risk factors for clinical decline. We recommend that all clinical risk factors be evaluated. Those with only a single risk factor are less likely to develop respiratory compromise and can be observed and treated on the ward. Those with two or more risk factors should be considered for ICU care.
11. When the patient is admitted to the ICU, it is important to utilize the available resources such as respiratory therapy and intensive nursing care to provide timely care. This includes hourly assessment of the patient's pulmonary status and pain control. Incentive spirometry goals should be set at 15 mL/kg up to a threshold of 1500 ccs. Below this volume the patient is at risk for decline. In

addition, the patient should be encouraged to perform strong, productive coughing. The patient's pain control regimen should be escalated if ineffective at keeping pain scores below 7. After giving a maximum of 8 h for improvement of these parameters, if the patient remains below target thresholds, then invasive neuraxial treatment should be pursued either in the form of thoracic epidurals or paravertebral catheters [15, 16]. Earlier placement can result in faster control of pain; therefore, the process should not be delayed as determined by the clinician. Consider early (<72 h) surgical stabilization of rib fractures if patient has flail chest or severely displaced fractures with respiratory compromise (see Chap. 8).

12. When the patient is admitted to the ward, interventions by the respiratory therapist and nursing staff are usually available every 4 h. Using this time interval, the patient should be encouraged to perform the same pulmonary evaluation and maneuvers as in the ICU. Incentive spirometry goals should be set at 15 mL/kg up to a threshold of 1500 ccs. In addition, the patient should be encouraged to perform strong, productive coughing. The patient's pain control regimen should be escalated if ineffective at keeping pain scores below 7. After giving a maximum of 8 hours for improvement of these parameters, if the patient remains below target thresholds, then invasive neuraxial treatment should be pursued either in the form of thoracic epidural or paravertebral catheter [15, 16]. Should the patient exhibit respiratory decline, a transfer to the ICU is warranted. Similarly regarding SSRF, consider early (<72 h) surgical stabilization of rib fractures if patient has flail chest or severely displaced fractures with respiratory compromise (see Chap. 8).

References

1. Todd SR, McNally MM, Holcomb JB, Kozar RA, Kao LS, Gonzalez EA, Cocanour CS, Vercruysse GA, Lygas MH, Brasseaux BK, et al. A multidisciplinary clinical pathway decreases rib fracture-associated infectious morbidity and mortality in high-risk trauma patients. Am J Surg. 2006;192:806–11.
2. Holcomb JB, Moore FA, et al. Morbidity from Rib Fractures Increases After Age 45. JACS. 2003 Oct;196(4):549–55.
3. Dehghan N, de Mestral C, McKee MD, Schemitsch EH, Nathens A. Flail chest injuries: a review of outcomes and treatment practices from the National Trauma Data Bank. J Trauma Acute Care Surg. 2014;76:462–8.
4. Hamrick MC, Duhn RD, Ochsner MG. Critical evaluation of pulmonary contusion in the early post-traumatic period: risk of assisted ventilation. Am Surg. 2009;75:1054–8.
5. Witt C, Bulger E. Comprehensive approach to the management of the patient with multiple rib fractures: A review and introduction of a bundled rib fracture management protocol. Trauma Surgery Acute Care Open. 2017;2:e00064.
6. Bulger EM, Arneson MA, Mock CN, Jurkovich GJ. Rib fractures in the elderly. J Trauma. 2000;48:1040–6. discussion 1046–7
7. Stawicki SP, Grossman MD, Hoey BA, Miller DL, Reed JF III. Rib fractures in the elderly: a marker of injury severity. J Am Geriatr Soc. 2004;52:805–8.
8. Rockwood K, et al. A global clinical measure of fitness and frailty in elderly people. CMAJ. 2005 Aug 30;173(5):489–95.

9. Brasel K, Moore E, Biffl W, et al. Western Trauma Association Critical Decisions in Trauma: Management of Rib Fractures. J Trauma Acute Care Surgery. 2016;82(1):200.
10. Battle CE, Hutchings H, Evans PA. Risk factors that predict mortality in patients with blunt chest wall trauma: a systematic review and meta-analysis. Injury. 2012;43:8–17.
11. Galvagno SM Jr, Smith CE, Varon AJ, Hasenboehler EA, Sultan S, Shaefer G, To KB, Fox A, Alley DE, Ditillo M, et al. Pain management for blunt thoracic trauma: a joint practice management guideline from the Eastern Association for the Surgery of Trauma and Trauma Anesthesiology Society. J Trauma Acute Care Surg. 2016;81:936–51.
12. Rafiq S, Steinbruchel DA, Wanscher MJ, Andersen LW, Navne A, Lilleoer NB, Olsen PS. Multimodal analgesia versus traditional opiate based analgesia after cardiac surgery, a randomized controlled trial. J Cardiothorac Surg. 2014;9:52.
13. Bayouth L, Safcsak K, Cheatham ML, Smith CP, Birrer KL, Promes JT. Early intravenous ibuprofen decreases narcotic requirement and length of stay after traumatic rib fracture. Am Surg. 2013;79:1207–12.
14. Yang Y, Young JB, Schermer CR, Utter GH. Use of ketorolac is associated with decreased pneumonia following rib fractures. Am J Surg. 2014;207:566–72.
15. Bulger EM, Edwards T, Klotz P, Jurkovich GJ. Epidural analgesia improves outcome after multiple rib fractures. Surgery. 2004;136:426–30.
16. Mohta M, Verma P, Saxena AK, Sethi AK, Tyagi A, Girotra G. Prospective, randomized comparison of continuous thoracic epidural and thoracic paravertebral infusion in patients with unilateral multiple fractured ribs—a pilot study. J Trauma. 2009;66:1096–101.

Non-union

11

John G. Edwards and William J. Hunt

Definition of Non-united Rib Fractures

Non-union of long bone fractures causes a significant challenge in trauma surgery. However, in the same way that the potential for the surgical management of multiple rib fractures is often overlooked, the clinical impact of non-united rib fractures is poorly appreciated. Although the long-term quality-of-life impact of multiple rib fractures has been reported [1, 2], the potential contribution of non-united rib fractures [NURFs] to the prolonged morbidity described is less well understood. An acceptable definition of a NURF is symptomatic non-union at 3 or more months post-injury [3]. Typical indications for consideration of surgical management include significant pain or instability attributable to the non-union(s), as visualised radiologically, occurring 3 months or more post-injury. The use of 3D composite volume rendered computed tomography (CT) scans is useful in determining the site and nature of NURFs (Fig. 11.1).

History

The first report of surgical stabilization of non-united rib fractures (SSNURFs) was published by Leavitt in 1942 [4]. He describes a patient suffering three painful NURFs dating from 1937. A year after the injury, he performed surgery which involved debridement and splitting each end of the involved ribs longitudinally in a

J. G. Edwards (✉) · W. J. Hunt
Department of Cardiothoracic Surgery, Northern General Hospital, Sheffield Teaching Hospitals NHS Foundation Trust, Sheffield, South Yorkshire, UK
e-mail: william.hunt@sth.nhs.uk; john.edwards@sth.nhs.uk

M. de Moya, J. Mayberry (eds.), *Rib Fracture Management*,
https://doi.org/10.1007/978-3-319-91644-6_11

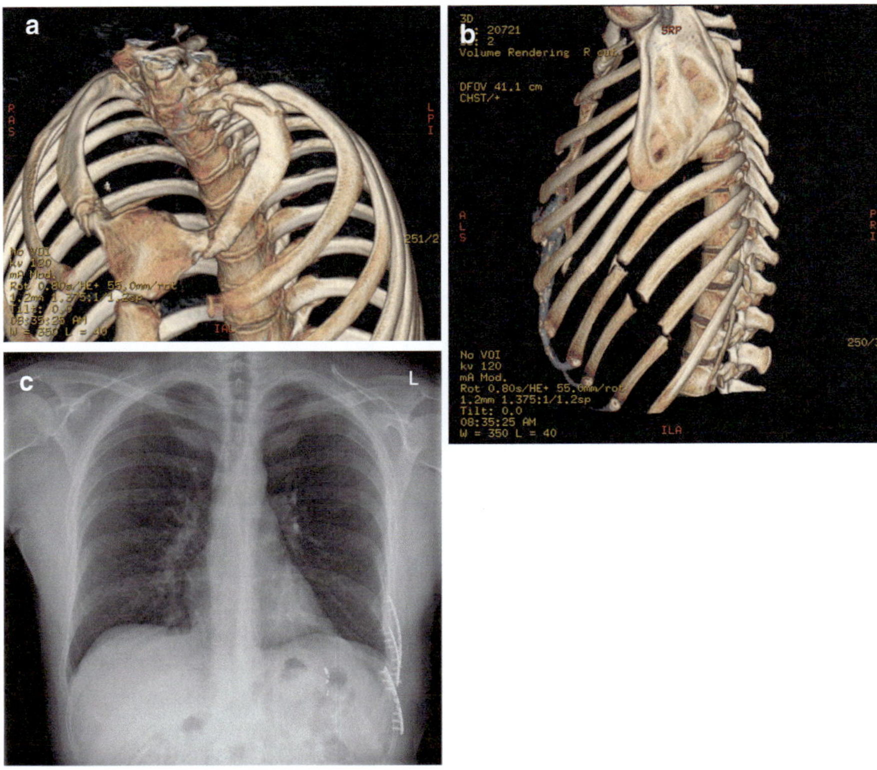

Fig. 11.1 A 34-year-old male, injured in a high-velocity road traffic collision. (**a**) Three-dimensional CT scan reconstructions of the ribs 4 months after the injury, showing non-union of a left first rib fracture. (**b**) A series of painful, atrophic, non-united fractures of the left eighth, ninth and tenth ribs. The seventh rib fracture has united. (**c**) Chest radiograph 6 weeks after surgical stabilisation of the eighth, ninth and tenth non-united rib fractures, with debridement of the rib ends, iliac crest bone grafting and locking plates and screws. Note the splenic artery embolisation coil and that one screw has loosened

vertical plane with an osteotome. Cortical bone grafts harvested from the tibia were shaped into three shuttles and driven into the split rib ends on each side of each non-union. A year later, reoperation was required, due to the formation of a painful prominence over the end of one graft. One rib fracture was noted to have united. The shuttle graft of another was united to the rib on one side of the non-union but protruding over the other fragment. This was debrided and packed with cancellous bone chips from the same rib and fixed with wire cerclage. The third fracture had not united and was also debrided and packed with bone chips. A plaster cast was applied to the trunk of the patient for 12 months. Symptoms were then reported to have improved significantly. Although the practice of surgical stabilization of rib fractures grew over the years [5], it is noteworthy that there were no subsequent reports of SSNURFs in the literature for 50 years.

Bone-Healing and Non-union Theory (BHN Theory)

The majority of non-unions will heal if the correct mechanical environment is produced by surgery.—D. S. Elliott, 2016.

Elliott and colleagues combined Wolff's law, Perren's strain theory and Frost's concept of the mechanostat to produce a unified theory of bone healing and non-union that can be applied to explain non-united rib fractures and SSNURF techniques [6]. The granulation tissue that forms after a fracture progressively stiffens, reducing the strain (the change in length of the tissues according to load) until it is low enough for bone formation. If mobility of the healing fracture exceeds the strain tolerance of the bone-healing unit, then the fracture will not heal. Creation of a mechanical stable environment, such that the strain tolerance is in the range whereby the BHU can function, is proposed as the key to the union of non-united fractures.

Different scenarios of excessive movement result in different types of non-union, depending on the strain of the BHU and the strain tolerance of the tissue within the BHU. According to Weber and Czech, hypertrophic, atrophic and infective types of non-union may occur [7]. Although the first two are commonly seen in thoracic trauma (Fig. 11.2), infective non-united rib fractures have not been reported. There are also different proposed aetiologies of non-union. The avascular model proposes that devascularisation occurs in atrophic non-union: however, there has been a

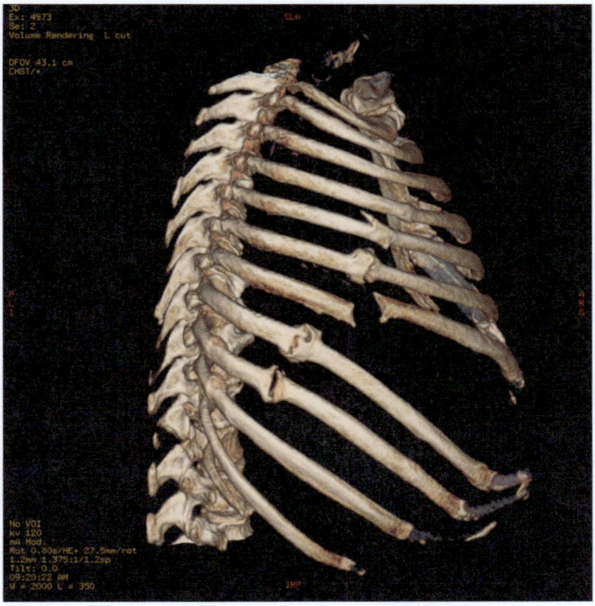

Fig. 11.2 Non-union of cough-induced rib fractures in a 61-year-old male. Hypertrophic non-united rib fractures (NURFS) of the right sixth, eighth and ninth ribs are seen: the seventh is of the atrophic type. This patient also had a costal margin rupture with intercostal hernia

failure experimentally to demonstrate this. It is thus proposed that the primary aetiology of non-union is mechanical [6]. Elliott proposes that there is an intact BHU in most non-unions but that the failure of the BHU is due to mechanical conditions. This can be either because fixation fails to reduce strain to a level at which new bone formation can occur to bridge the fracture or because the fixation construct is so stiff that net bone resorption occurs. In the former scenario, if excessive strain persists at the fracture site, movement prevents the normal function of the bone-healing unit, resulting in the formation of a pseudoarthrosis (Fig. 11.2). Furthermore, even with plate fixation, if anatomical reduction is not achieved, a gap may remain leaving a situation with persistently high strain, with excessive movement still occurring, such that the result is either the fracture of the plate or loosening of the screws (Fig. 11.3).

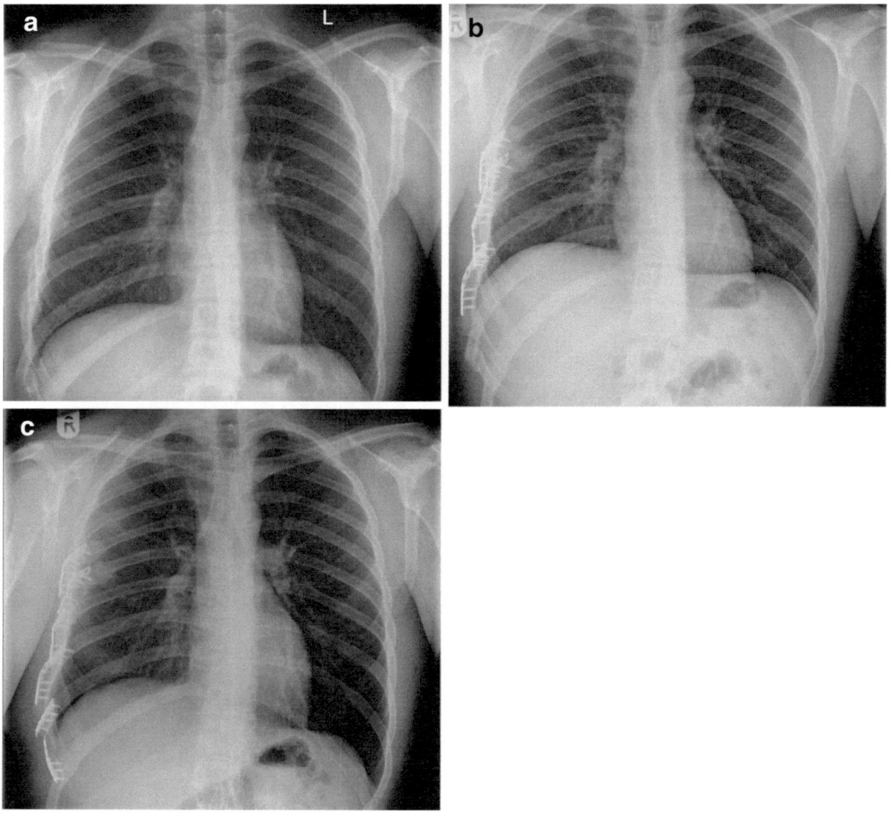

Fig. 11.3 A 21-year-old male tennis professional with spontaneous right sixth–eleventh NURFs. (**a**) Chest radiograph demonstrating hypertrophic NURFs. (**b**) Chest radiograph 3 months after SSNURFs of the right. The tenth rib fracture felt united and was not stabilised. However, pain localised to this persisted, and he underwent further surgery. (**c**) Plating of the tenth NURF was performed after excision of the non-union, leaving a 1 cm gap which was not bone grafted. The titanium plate fractured 3 weeks postoperatively

It is recognised that there are biological influences, which may affect the healing of fractures and the incidence of non-union. These include smoking, endocrine diseases (including but not exclusively diabetes mellitus), prior radiotherapy, peripheral vascular disease and the severity and location of the fracture [8]. These biological factors may reduce fracture healing through the delay of prevention of new bone formation or narrowing the strain range where new bone formation can occur. With regard to rib fractures, the impact of these biological influences is unknown, as is the role of modulation of them in the management of non-united fractures.

Hence, Elliot proposes that treatment of a non-union will be successful if strain at the site of the non-union is reduced to lie within the range at which bone formation can occur [6]. This can either be achieved by reducing the forces applied across the fracture or by direct stabilization. With regard to the former method, it is interesting to note the common practice in a previous era of "strapping" ribs, or even the application of a trunk plaster cast, as performed by Leavitt [4]. This fell out of practice over the decades, although the use of corset-like devices has come into fashion in cardiac surgery, for patients where there is a high risk of non-union of the sternotomy [9]. Elliot suggests that there are three central aspects to the management of non-union. Firstly, there is no requirement to excise tissue from the fracture site, which will heal if the mechanical environment is corrected. Secondly, when a low strain environment can be achieved, union will occur without the use of bone morphogenetic proteins (BMPs) [10]. Thirdly, with acknowledgement to the principles of BHN theory, it is only necessary to consider the use of biologic adjuncts, including bone grafting, when there is significant bone loss.

Hence most, if not all, non-united rib fractures are due to excessive strain. Application of BHN theory to non-united rib fractures suggests that:

1. It is mechanical environment that is the key to ensuring union.
2. External cortical plating should be successful, with minimal or no fracture debridement.
3. BMPs are not required to gain union.
4. Bone grafting is only required if metalwork has to bridge a significant gap, whereby there may be greater risk of metalwork failure.

Literature Review

With understanding of BHN theory in mind, the literature is reviewed. The different aspects of surgical technique (management of the non-union, whether and how bone grafting is performed and the technique of fixation) are indicated in Table 11.1. There are a number of case reports in the literature (Table 11.2), which show variation in the techniques used [11–23]. However, there are few objective measures of the outcomes and a lack of long-term follow-up, in order to determine symptomatic and radiographic response to the SSNURF.

Table 11.1 Different aspects of surgical technique for surgical stabilisation of non-united rib fractures (SSNURFS)

Non-union	Bone graft	Bone graft source	Plates	Plate fixation
Nothing	None	Hypertrophic non-union	None	Screws
Debride	Intramedullary shuttle with the cortical bone	Iliac crest	Titanium	Wire circlage
Debride and ream	Inlay graft with cortical bone	Neighbouring rib(s)	Absorbable	
Resect	Cortical/cancellous chips	Tibia		

The Oregon Health and Science University Experience

Recognising this, Fabricant and colleagues performed a prospective clinical study [3] in 24 patients, presenting over nearly 4 years (many from out of state). In order to assess patient-reported outcome measures, they used the McGill Pain Questionnaire (MPQ) and SF-36 Quality-of- Life Questionnaire, administered before the operation and at 2, 4 and 6 months postoperatively. Their surgical technique involved resection of the non-union and then assessment of the width of the gap between the cut ends of the rib. A gap less than 1 cm was plated, with a titanium plate and screws if the non-union was situated posterolaterally or was displaced, and an absorbable plate fixed with absorbable suture cerclage was used for anterior non-unions where there was minimal or no displacement. Gaps of greater than 2 cm following fixation were not usually fixated, but care was taken to ensure that non-united fractures on neighbouring ribs were not resected in order to reduce the chance of non-union. Occasionally, however, absorbable plates and cerclage were used. For fractures between 1 and 2 cm, the choice of technique was left to the surgeon. Where intercostal nerve entrapment was identified, steps were taken to free the nerve: this is a step that has been poorly addressed in other reports. Fabricant and colleagues noted that there was significant improvement at 6 months following surgery in many of the subjective outcome measures, such as the MPQ Present Pain Intensity and Pain Rating Index, as well as several domains of the SF-36 (physical functioning, role physical, social functioning, role social, bodily pain, vitality, mental health and general health). However, functional and work status did not improve, nor did the proportion of patients taking opioid analgesia. Complications were uncommon (one wound infection, two partial screw backouts, one chest wall hernia).

The University of Minnesota Experience

Gauger and colleagues identified ten patients (with a total of 16 NURFs) in 5.5 years, defining non-union as painful lack of union occurring on radiographs at least 3 months apart (mean 24 months from injury (range 8–60 months) [24]. The indication for surgery was continuous disabling pain, which was refractory to conservative management. The preferred surgical technique described included correction of the deformity and recontouring of the rib, with clearance of fibrous tissue

Table 11.2 Summary of published case reports of surgical stabilisation of non-united rib fractures (SSNURFs), indicating the surgical techniques used[a]

Author	Number of patients	Surgical technique				
		Non-union	Bone graft type	Bone graft source	Plates and fixation	Number of ribs treated
Leavitt [4]	1	Debride and ream	Intramedullary tibial cortical shuttle	Tibia	None	3
Reber [11]	1	Not stated	None	None	Plates and screws	2
Morgan-Jones [12]	1	Debride (longitudinal gutter)	Cortical inlay	Iliac crest	None	1
Cacchione [13]	1	Debride	None	None	Plates and screws	3
Slater [14]	1	Excision	None	None	Plates and wire cerclage	2
Ng [15]	1	Debride	None	None	Plates and screws	3
Gardenbroek [16]	3	Not stated	None	None	Locking plates and screws	1, 3, 3
Cho [17]	1	Excision	Inlay block	Tibia	None	1
Anavian [18]	1	Debride and ream	Chips	Hypertrophic non-union	Locking plates and screws	3
Dean [19]	1	Debride	Bone chips	Iliac crest	Plates and screws	3
Kaplan [20]	1	Debride	None	None	Plates and screws	3
Proffer [21]	1	Excision	None	None	None	1
Puttman [22]	1	Excision	None	None	None	1
Terabayahsi [23]	1	Excision	None	None	None	1

[a]The series of Fabricant ($n = 24$) [3] and Gauger ($n = 10$) [24] are discussed in detail in the text

connections from the ends of the non-union; opening the medullary canals with the use of a 3.5 mm drill bit in oscillate mode; fixation with external cortical locking plates; and autograft bone grafting with excess local bone or iliac crest bone grafting with either cancellous chips or a tricortical graft for longer defects. Between 1 and 3 (median 2) NURFs were reconstructed using these techniques. Union was noted on postoperative chest radiographs. Preoperative SF-36 scores were not recorded: postoperatively, the mean mental and physical component scores were 54.4 and 43.5, respectively; eight out of the ten patients returned to work and/or their previous activities without any limitations.

The Sheffield Experience

We commenced a programme of surgical stabilisation of rib fractures in 2006. Between 2009 and 2017, we performed 26 operations on 24 patients with 63 NURFs (median 3 (range 1–6) NURFs fixed per patient). Seventeen (71%) were male, the median age was 57 (range 24–75) years, and interval from surgery median is 6.3 (range 3–69) months. On review of the CT scans dating from the time of trauma, only six patients were judged to have a flail segment, but 83% were classified as having a displaced series of fractures (where a series is defined as fractures in a similar location on three or more successive ribs). One hundred and seventy-seven fractures were identified on the immediate post-injury CT scan, but 87 (49%) were judged not to have united at the time of surgery, and 63 (36%) NURFs were stabilised surgically. Indications were painful NURFs in 21 (88%) patients and deformity in one patient.

The surgical techniques we have used, summarised in Table 11.3, are consistent with BHN theory, as described earlier. Minimal debridement of the NURF is performed, with the primary aim of recontouring the external cortex of a hypertrophic non-union to provide an even curved surface for external cortical plating. For atrophic NURFs, resection of the tips is performed to reveal cancellous bone, although reaming of the intramedullary cavity is not performed. Excision of the non-union was only performed in one patient with grossly hypertrophic NURFs. Bone grafting, usually from bone chips harvested during the recontouring, is reserved for atrophic NURFs, when reduction to create contact between the bone fragments cannot be achieved. Harvesting cortical-cancellous bone graft from the iliac crest was performed in one patient. External cortical plating was performed with

Table 11.3 The Sheffield series of surgical stabilisation of non-united rib fractures

Non-union		Bone graft		Bone graft source		Plates		Plate fixation	
No debridement	19	None	45	Hypertrophic non-union	15	None	1	Screws	62
Debride	42	Intramedullary shuttle with cortical bone	0	Iliac crest	3	Titanium	62	Wire cerclage	0
Debride and ream	0	Inlay graft with cortical bone	0	Neighbouring rib(s)	0	Absorbable	0		
Resect	2	Cortical/cancellous chips	18	Tibia	0				

The frequencies are given of the different aspects of surgical management for each of the 63 NURFs, which occurred in a total of 24 patients titanium reconstruction ribbon (Depuy Synthes) in six patients, prior to 2011, since when the MatrixRib system (Synthes, West Chester PA, USA) has been used.

There was one case of titanium plate fracture, where the plate was used to bridge an excised NURF, leaving a 1 cm gap that was not packed with bone graft chips (Fig. 11.3c). However, the patient has declined to have the metalwork removed. A loosened screw was noted in another patient (Fig. 11.1c), but this did not require removal as the patient was well. There were no postoperative wound infections or other adverse events following surgery. We did not perform CT scans routinely post-operatively to confirm union. All patients were followed up in the clinic and underwent chest radiography.

We sent postal quality-of-life questionnaires (SF-26, Brief Pain Inventory, modified Glasgow Outcome Scale extended) to all patients. There were 11 responses (46%), and hence it is accepted that there is potential bias from the small number of responses. Preoperative symptoms were reported as severe pain in seven (64%) and moderate pain in three (27%) patients. There was dyspnoea in three patients, and one reported their sleep being affected prior to surgery. Following surgery, seven (64%) reported that their pain had not resolved completely, whereas four (36%) had no pain whatsoever. Two patients had persistent pain which was determined to originate from NURFs distant to those that were stabilised at surgery: these underwent a second operation. When asked if patients thought that SSNURFs was the correct course of management, eight (73%) said yes, with two of those patients reporting that the surgery was positively life-changing. However, three (27%) were unsure if surgery was the correct option.

Non-union of First Rib Fractures

Non-union of first rib fractures (Fig. 11.1a) can cause not only persistent pain [21] but also thoracic outlet syndrome (TOS) [22, 23, 25, 26]. There are multiple reports of non-united first rib fractures occurring in athletes. When persistent pain or TOS is present, standard resection of the first rib, for example, by the transaxillary approach, may result in resolution of symptoms. No bone grafting or plating need be performed, although fractures near the first costochondral junction may be plated across the fracture from the rib to the manubrium [T. W. White, personal communication].

Nonoperative Management of Non-united Rib Fractures

Many risk factors for the generation of non-union long bone fractures have been proposed [27, 28]. However, where there is significant beneficial role of risk factor modulation in the successful management of fracture, non-union is subject to much debate. There are no specific data regarding rib fractures. Of course, optimisation of concurrent diseases is appropriate. Smokers have a greater risk of non-union of long bones [Pearson], but it is unknown whether smoking cessation results in an increased rate of union after surgery. Some surgeons will insist in smoking cessation prior to rib fracture non-union surgery [3]. Measurement of vitamin D levels is advocated by some, although vitamin D supplementation did not increase the already high union rates in a randomised trial of patients with hypovitaminosis D suffering from long bone fractures [29].

The use of healing adjuncts is also subject to much debate. In some healthcare systems, techniques such as low-intensity pulsed ultrasound (LIPUS), electromagnetic stimulation (ESTIM) [30] and low-level laser therapy (LLLT) [31] are available in an attempt to aid the healing of acute and non-united fractures. LIPUS is being used in some centres for the management of rib fractures. Importantly, whilst there are anecdotal reports of success for rib fractures, there are no published data or clinical trials to demonstrate efficacy or cost-effectiveness. It is interesting to note that Gauger used a trial of a bone stimulator in nine of the ten patients undergoing SSNURFs, for at least a 3-month duration in each case, but in only one patient did one of the three NURFs unite [24].

A Cochrane systematic review in 2012 noted methodological deficiencies in clinical trials and found insufficient evidence to recommend routine use in clinical practice [32]. There have been mixed reports more since: a recent clinical trial have failed to find better tibial fracture healing with LIPUS [33]. A recent meta-analysis noted that LIPUS treatment reduces the time to fracture union but does not reduce the rate of non-union [34]. In another meta-analysis, in patients with an existing non-union or delayed union, ESTIM appeared to have a benefit over standard care on union rates at 3 months [35]. However, there are randomised trials demonstrating both benefit and no benefit [36, 37], and hence any recommendations are to be taken with caution. Clinical trials of such technologies are required to determine whether there are benefits in the management of acute and non-united rib fractures.

Conclusions

Although the benefits of SSNURFs are clear, the evidence is limited to case reports and small cohort studies. Further prospective clinical trials are required, with objective as well as subjective outcome measures in order to define clearly the benefits of surgical management. In the absence of a large robust, prospective database with comprehensive follow-up directed at the assessment of union of each fracture, it is difficult to be able to assess the true incidence and clinical and patient impact of NURFs. Such work would also help to determine predictive factors for the generation of NURFs, which would allow refinement of the indications for and techniques of SSRF [38–40]. Clinical trials are also required to determine the optimum surgical strategy and whether there is value to nonoperative strategies such as the external bone stimulation technologies.

References

1. Fabricant L, Ham B, Mullins R, Mayberry J. Prolonged pain and disability are common after rib fractures. Am J Surg. 2013;205(5):511–5.

2. Marasco S, Lee G, Summerhayes R, Fitzgerald M, Bailey M. Quality of life after major trauma with multiple rib fractures. Injury. 2015;46(1):61–5.
3. Fabricant L, Ham B, Mullins R, Mayberry J. Prospective clinical trial of surgical intervention for painful rib fracture nonunion. Am Surg. 2014;80(6):580–6.
4. Leavitt DG. Non-union of three ribs. J Bone Joint Surg. 1942;1942:932–6.
5. Bemelman M, Poeze M, Blokhuis TJ, Leenen LP. Historic overview of treatment techniques for rib fractures and flail chest. Eur J Trauma Emerg Surg. 2010;36(5):407–15.
6. Elliott DS, Newman KJH, Forward DP, Hahn DM, Ollivere B, Kojima K, et al. A unified theory of bone healing and nonunion: BHN theory. Bone Joint J. 2016;98-B(7):884.
7. Weber BG, Czech O. Pseudarthrosen–pathophysiologie biomechanik. Huber: Therapie Ergebnisse; 1973.
8. Gaston MS, Simpson AH. Inhibition of fracture healing. J Bone Joint Surg Br. 2007;89(12):1553–60.
9. Gorlitzer M, Wagner F, Pfeiffer S, Folkmann S, Meinhart J, Fischlein T, Reichenspurner H, Grabenwoeger M. Prevention of sternal wound complications after sternotomy: results of a large prospective randomized multicentre trial. Interact Cardiovasc Thorac Surg. 2013;17(3):515–22.
10. Barcak EA, Beebe MJ. Bone morphogenetic protein: is there still a role in orthopedic trauma in 2017? Orthop Clin North Am. 2017;48(3):301–9.
11. Reber P, Ris HB, Inderbitzi R, Stark B, Nachbur B. Osteosynthesis of the injured chest-wall – use of the AO (arbeitsgemeinschaft fur osteosynthese) technique. Scand J Thorac Cardiovasc Surg. 1993;27(3–4):137–42.
12. Morgan Jones RL, Mackie IG. Non-union of a fracture of the 8th rib. Injury. 1996;27(2):147–8.
13. Cacchione RN, Richardson JD, Seligson D. Painful nonunion of multiple rib fractures managed by operative stabilization. J Trauma Injury Infect Crit Care. 2000;48(2):319–21.
14. Slater MS, Mayberry JC, Trunkey DD. Operative stabilization of a flail chest six years after injury. Ann Thorac Surg. 2001;72(2):600–1.
15. Ng ABY, Giannoudis PV, Bismil Q, Hinsche AF, Smith RM. Operative stabilisation of painful non-united multiple rib fractures. Injury. 2001;32(8):637–9.
16. Gardenbroek TJ, Bemelman M, Leenen LPH. Pseudarthrosis of the ribs treated with a locking compression plate a report of three cases. J Bone Joint Surg. 2009;91A(6):1477–9.
17. Cho YH, Kim HK, Kang D-Y, Choi YH. Reoperative surgical stabilization of a painful non-united rib fracture using bone grafting and a metal plate. J Orthop Trauma. 2009;23(8):605–6.
18. Anavian J, Guthrie ST, Cole PA. Surgical management of multiple painful rib nonunions in patient with a history of severe shoulder girdle trauma: a case report and literature review. J Orthop Trauma. 2009;23(8):600–4.
19. Dean NC, Van Boerum DH, Liou TG. Rib plating of acute and sub-acute non-union rib fractures in an adult with cystic fibrosis: a case report. BMC Res Notes. 2014;7:681.
20. Kaplan DJ, Begly J, Tejwani N. Multiple rib nonunion: open reduction and internal fixation and iliac crest bone graft aspirate. J Orthop Trauma. 2017;31:S34–5.
21. DS P, Patton JJ, Jackson DW. Nonunion of a first rib fracture in a gymnast. Am J Sports Med. 1991;19(2):198–201.
22. Puttmann KT, Satiani B, Vaccaro P. Thoracic outlet syndrome in a volleyball player due to nonunion of the first rib fracture. Vasc Endovasc Surg. 2016;50(8):563–5.
23. Terabayashi N, Ohno T, Nishimoto Y, Oshima K, Takigami I, Yasufuku Y, Shimizu K. Nonunion of a first rib fracture causing thoracic outlet syndrome in a basketball player: a case report. J Shoulder Elb Surg. 2010;19(6):e20–3.
24. Gauger EM, Hill BW, Lafferty PM, Cole PA. Outcomes after operative management of symptomatic rib nonunion. Journal of orthopaedic trauma. J Orthop Trauma. 2015;29(6):283–9.
25. Young BL, Watson SL, Meyer RD, Ponce BA. Nonunion of first rib fracture in a softball player: case report of a rare cause of thoracic outlet syndrome. J Shoulder Elb Surg. 2016;25(11):e353–7.
26. Duane TM, O'Connor JV, Scalea TM. Thoracic outlet syndrome resulting from first rib fracture. J Trauma. 2007;62(1):231–3.

27. Zura R, Xiong Z, Einhorn T, Watson JT, Ostrum RF, Prayson MJ, Della Rocca GJ, Mehta S, McKinley T, Wang Z, Steen RG. Epidemiology of fracture nonunion in 18 human bones. JAMA Surg. 2016;151(11):e162775.
28. Richards CJ, Graf KW Jr, Mashru RP. The effect of opioids, alcohol, and nonsteroidal anti-inflammatory drugs on fracture union. Orthop Clin North Am. 2017;48(4):433–43.
29. Haines N, Kempton LB, Seymour RB, Bosse MJ, Churchill C, Hand K, Hsu JR, Keil D, Kellam J, Rozario N, Sims S, Karunakar MA. The effect of a single early high-dose vitamin D supplement on fracture union in patients with hypovitaminosis D: a prospective randomised trial. Bone Joint J. 2017;99-B(11):1520–5.
30. Griffin XL, Costa ML, Parsons N, Smith N. Electromagnetic field stimulation for treating delayed union or non-union of long bone fractures in adults. Cochrane Database Syst Rev. 2011;4:CD008471. https://doi.org/10.1002/14651858.CD008471.pub2.
31. Iryanov YM. Influence of laser irradiation low intensity on reparative osteogenesis and angiogenesis under transosseous osteosynthesis. J Lasers Med Sci. 2016;7(3):134–8.
32. Griffin XL, Smith N, Parsons N, Costa ML. Ultrasound and shockwave therapy for acute fractures in adults. Cochrane Database Syst Rev. 2012;2:CD008579. https://doi.org/10.1002/14651858.CD008579.pub2.
33. TRUST Investigators Writing Group, Busse JW, Bhandari M, Einhorn TA, Schemitsch E, Heckman JD, Tornetta P III, Leung KS, Heels-Ansdell D, Makosso-Kallyth S, Della Rocca GJ, Jones CB, Guyatt GH. Re-evaluation of low intensity pulsed ultrasound in treatment of tibial fractures (TRUST): randomized clinical trial. BMJ. 2016;355:i5351. https://doi.org/10.1136/bmj.i5351.
34. Lou S, Lv H, Li Z, Zhang L, Tang P. The effects of low-intensity pulsed ultrasound on fresh fracture: a meta-analysis. Medicine (Baltimore). 2017;96(39):e8181.
35. Ebrahim S, Mollon B, Bance S, Busse JW, Bhandari M. Low-intensity pulsed ultrasonography versus electrical stimulation for fracture healing: a systematic review and network meta-analysis. Can J Surg. 2014;57(3):E105–18.
36. Martinez-Rondanelli A, Martinez JP, Moncada ME, Manzi E, Pinedo CR, Cadavid H. Electromagnetic stimulation as coadjuvant in the healing of diaphyseal femoral fractures: a randomized controlled trial. Colomb Med (Cali). 2014;45(2):67–71.
37. Adie S, Harris IA, Naylor JM, Rae H, Dao A, Yong S, Ying V. Pulsed electromagnetic field stimulation for acute tibial shaft fractures: a multicenter, double-blind, randomized trial. J Bone Joint Surg Am. 2011;93(17):1569–76.
38. Pieracci FM, Majercik S, Ali-Osman F, Ang D, Doben A, Edwards JG, French B, Gasparri M, Marasco S, Minshall C, Sarani B, Tisol W, VanBoerum DH, White TW. Consensus statement: surgical stabilization of rib fractures rib fracture colloquium clinical practice guidelines. Injury. 2017;48(2):307–21.
39. Bemelman M, de Kruijf MW, van Baal M, Leenen L. Rib fractures: to fix or not to fix? An evidence-based algorithm. Korean J Thorac Cardiovasc Surg. 2017;50(4):229–34.
40. Marasco S, Liew S, Edwards E, Varma D, Summerhayes R. Analysis of bone healing in flail chest injury: do we need to fix both fractures per rib? J Trauma Acute Care Surg. 2014;77(3):452–8.

Operative Techniques

12

Andrew Doben and Thomas W. White

Positioning

There are three principal positions: supine, lateral decubitus, and prone. The choice of position is informed by fracture pattern and anticipated exposure. Preoperative imaging with CT scanning and 3D reformatting will greatly assist in determining the optimal position. The use of a dual lumen tube is not required. However, if there is a plan for thoracoscopic intervention, dual lumen tubes may provide better visualization and access to the hemithorax. For open-standard approaches, this technique is not frequently utilized.

Supine

Best for anterior and anterolateral fractures, this position is achieved with the patient flat on their back with the ipsilateral arm abducted to 90°. The use of a bump can assist in the exposure of more anterolateral fractures. These authors also recommend placing the arm board as high as possible on the bed rail, to allow adequate room for a self-retaining retraction system. These devices are often invaluable in achieving and maintaining optimal exposure. A typical anterolateral thoracotomy placed in the inframammary fold will expose the majority of fractures in this region. This incision can be easily extended across the sternal midline or lateral to the axilla if additional exposure is needed.

A. Doben (✉)
Division of Trauma, Acute Care Surgery and Surgical Critical Care, Department of Surgery, University of Massachusetts Medical School–Baystate, Springfield, MA, USA
e-mail: Andrew.Doben@baystatehealth.org

T. W. White
Intermountain Trauma and General Surgery, Murray, UT, USA
e-mail: Tom.White@imail.org

M. de Moya, J. Mayberry (eds.), *Rib Fracture Management*,
https://doi.org/10.1007/978-3-319-91644-6_12

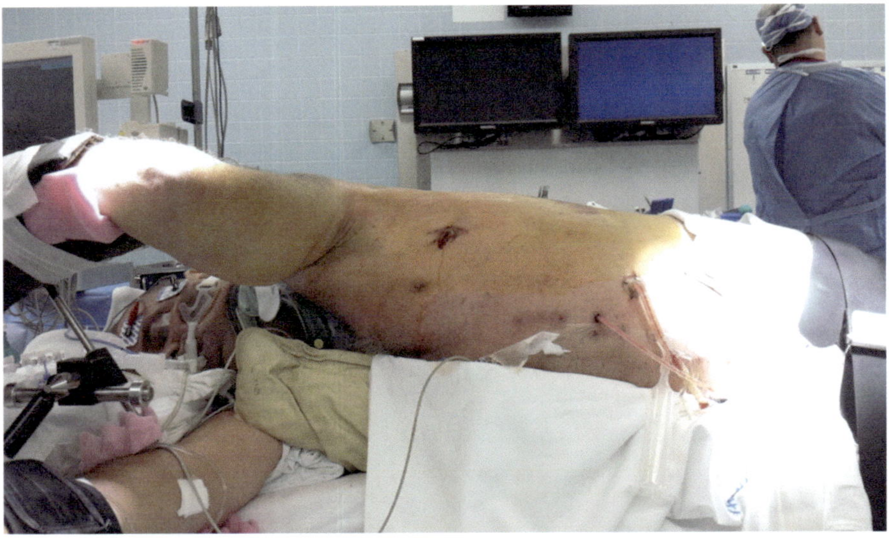

Fig. 12.1 Traditional lateral decubitus position

Lateral

Standard lateral decubitus positioning with a standard lateral thoracotomy is a versatile approach and should be considered as the default for most repairs (Fig. 12.1). This allows for exposure of the majority of unilateral fractures. When a combination of fracture lines (such as anterolateral on rib #4, posterolateral on rib #5, and both on rib #6) exists, this position should be the standard approach since it allows one to access more anterior and posterior exposures.

Prone

This positioning technique is well suited for posterior and posterolateral fractures. Face down, the patient is well padded, and the well-supported ipsilateral arm is allowed to hang over the side of the OR table. This allows for lateral and superior rotation of the scapula which will help to improve the exposure (Fig. 12.2).

Incisions

The incision can be made vertically or in a more standard posterolateral thoracotomy (Figs. 12.3, 12.4, and 12.5).

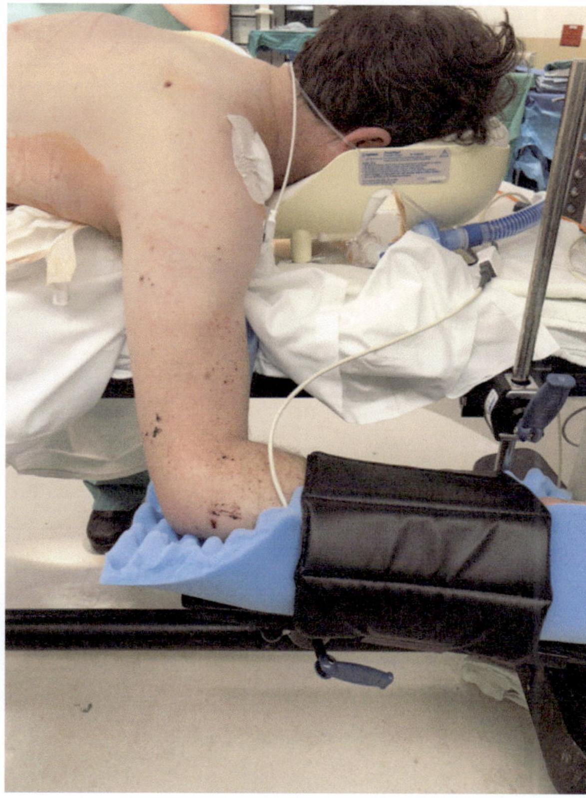

Fig. 12.2 Prone positioning with arm below table position to allow for lateral scapular rotation

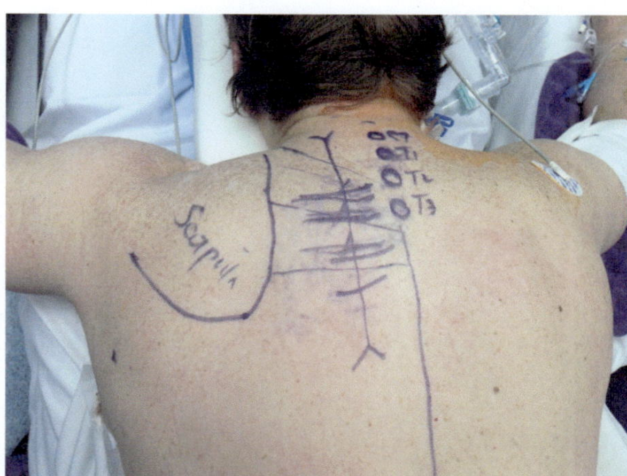

Fig. 12.3 Skin markings of posterior approach with patient positioned prone (Courtesy of Timothy H. Pohlman, MD, FACS, Indiana University Health)

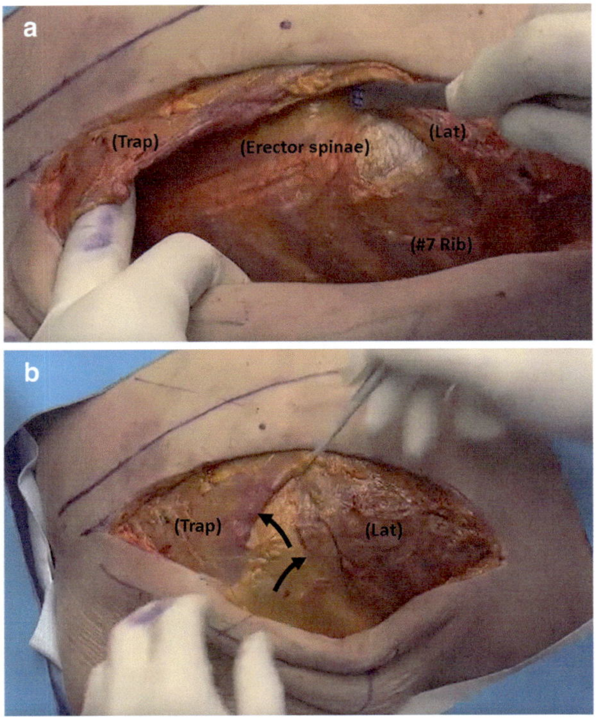

Fig. 12.4 (**a**, **b**) Dissection from a posterior approach through the auscultatory triangle (Courtesy of Timothy H. Pohlman, MD, FACS, Indiana University Health)

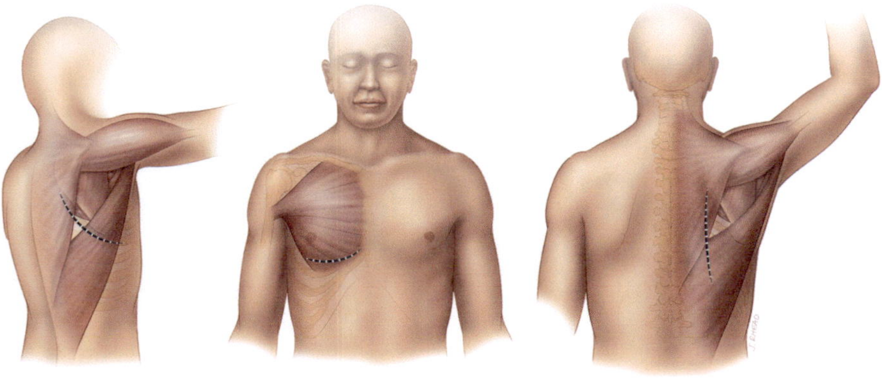

Fig. 12.5 The incision can be made vertically or in a more standard posterolateral thoracotomy (© 2017 Intermountain Healthcare. All Rights Reserved. Used with permission)

Exposure

After optimal positioning for a fracture pattern has been achieved, actual dissection to the level of the fractures becomes the next challenge. Subcutaneous flaps with muscle release can greatly facilitate muscle mobility and improve visualization (Figs. 12.6 and 12.7). The fractures are then palpated through the muscle to

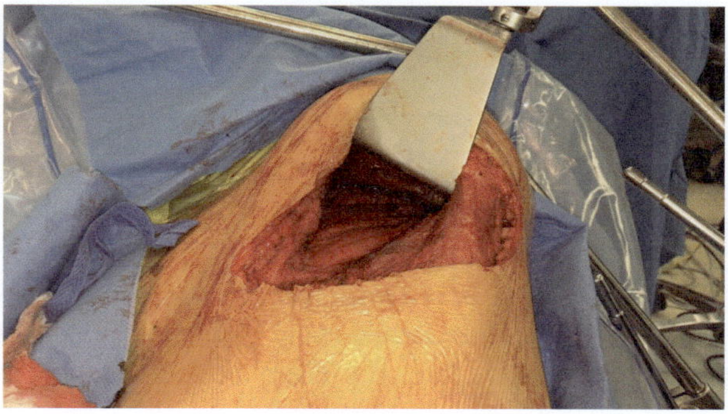

Fig. 12.6 Elevation of muscle flaps exposing the chest wall from a standard lateral thoracotomy approach

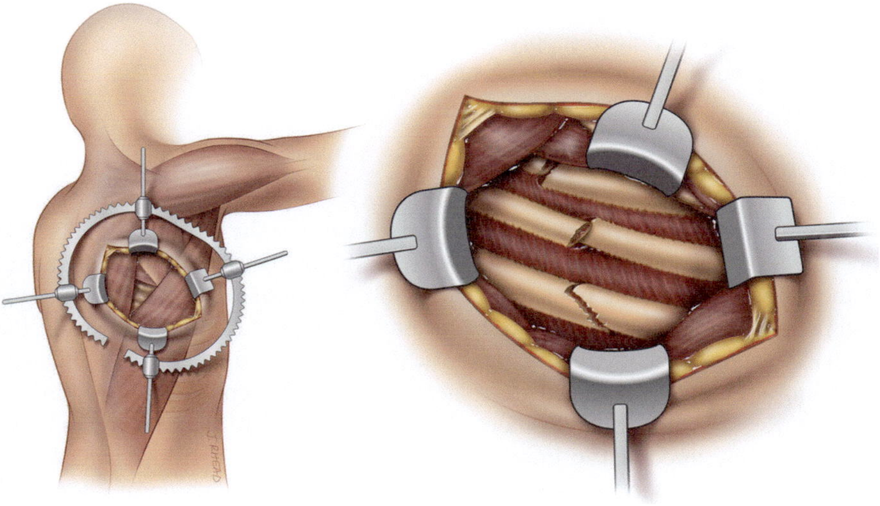

Fig. 12.7 Subcutaneous flaps with muscle release can greatly facilitate muscle mobility and improve visualization (© 2017 Intermountain Healthcare. All Rights Reserved. Used with permission)

determine the exact location. All fractures will require a minimum of 3 cm of exposure on either side of the fracture lines (which include both the inner and outer table of the fracture line) which can be quite offset in the complex fracture pattern.

Retraction

Once the fractures (both the inner and outer table) are identified, one begins to mobilize or split the muscle in the direction of its fibers. The goal is to avoid division of muscle if possible. Developing a soft tissue dissection envelope will allow for retraction of the musculature. We prefer to maintain this exposure with a self-retaining retraction system, such as the Bookwalter. The anterior surfaces of the fractured rib ends are then gently exposed. A vigorous removal of the periosteum is ill advised and unnecessary. It is helpful to enter the pleura at times to facilitate reduction of the fracture in a space above or below the fracture site.

Orthopedic Principles

The AO foundation was founded in 1958 as the *Arbeitsgemeinschaft für Osteosynthesefragen* (German for "Association for the Study of Internal Fixation"). The organization has created many standards on methods of bone plating and surgical treatment of fractures [1]. There are four AO principles:

1. Anatomic reduction
2. Stable fixation
3. Preservation of blood supply
4. Early mobilization

Reduction of Fracture

Of the four AO guiding principles, anatomic reduction is the most important intraoperative consideration. While stable fixation is critical (see Hardware Securement), without proper "anatomic reduction to reestablish anatomical relationships," optimal pulmonary physiology may not be restored. This is an important aspect of the technical approach to SSRF. This can be achieved through a variety of techniques, including specifically designed instrumentation from all commercially available systems. Restoring the natural contour of the chest, especially in the "stoved in" flail chest, is vitally important to improving postoperative pulmonary mechanics. The images below represent proper anatomic reduction and restoration of chest wall contour (Figs. 12.8 and 12.9). *Early intervention tends to make anatomic reduction easier* [2] *due to lack of more severe inflammation.* Fracture reduction should take place at each individual fracture site, and care should be made to determine the classification of the fracture to ensure that the entire fracture line has been reduced. It is important to recognize that the fracture line on the anterior cortex may not be the same as the fracture line on the posterior cortex. Once bicortical reduction and chest wall contour have been achieved, it is necessary to secure the anatomic alignment via a construct.

Hardware Securement

Hardware securement needs to achieve optimal positioning of the anatomic alignment without compromising the blood supply to the bone: AO principles 2 and 3. Several commercially available rib-specific systems use different methods and instrumentation to achieve this goal. Regardless of the system utilized, the primary goal is to temporarily secure the reduction and the fixation device while achieving

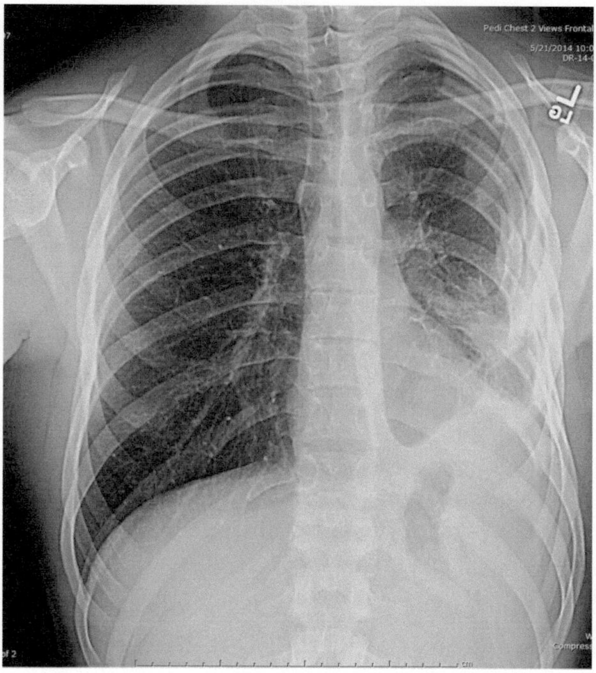

Fig. 12.8 Postoperative imaging demonstrating reduction of fractures and restoration of chest wall contour

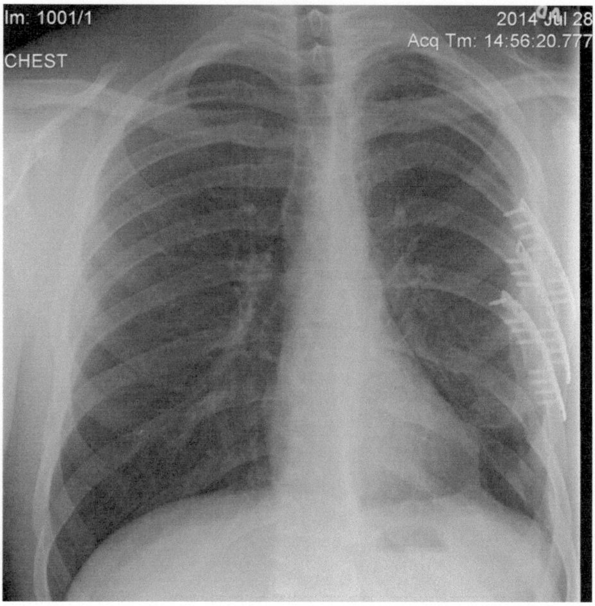

Fig. 12.9 Preoperative imaging demonstrating displacement and reduction of volume of the left hemithorax

permanent fixation with proper hardware. This can be done with clamping instrumentation or temporary fixation devices such as screws and threaded reduction tools (Fig. 12.10).

Primary goals of this portion of the operation are to maintain anatomic rib alignment and to ensure that the rigid fixation "plate" and screw system achieves sturdy periosteal apposition to the bone to reduce the risk of plate migration or "pull out." All of the available systems utilize locking screw technology which strengthens the entire construct and minimizes failure (Fig. 12.11).

While these authors recommend fixing as many locations as feasible through the original incision, not all fractures require repair. Typically ribs 1, 2, 11, and 12 do not require repair. Ribs 3 through 10 are repaired when necessary to achieve chest wall stability and to maintain contour. When "flail segments" or floating segments (two fractures on the same rib) exist, it has been our practice to fix both fractures as not doing this has been shown to not result in return of anatomic contour [3].

Because each level of fixation will incrementally enhance the contour and stability of the chest wall, fractures that cannot be approached without undue morbidity and do not contribute to overall instability do not require fixation.

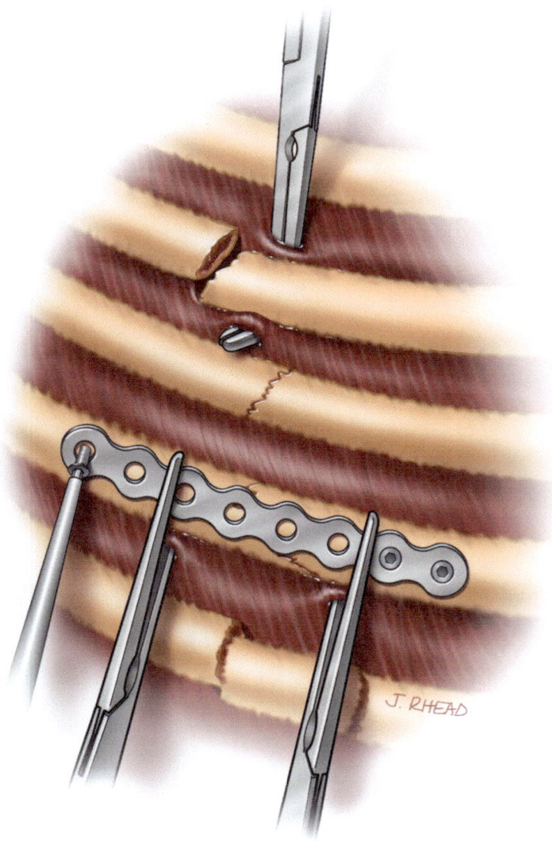

Fig. 12.10 Temporary fixation devices (© 2017 Intermountain Healthcare. All Rights Reserved. Used with permission)

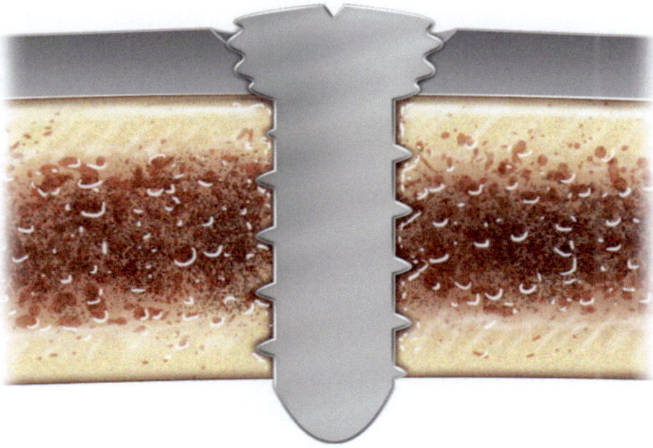

Fig. 12.11 Locking screw technology strengthens the entire construct and minimizes failure (© 2017 Intermountain Healthcare. All Rights Reserved. Used with permission)

Closure

Multiple Layers

Proper closure technique helps return vascularity to the dissected space as well as minimizes potential space for fluid collections. Muscle sparing and splitting techniques result in minimal residual anatomic disruption. These authors use absorbable suture to re-approximate musculature to the anatomic position and minimize potential space. The soft tissue layers are closed in multiple layers without tension to create natural planes separating the skin from the deep hardware in an attempt to minimize postoperative hardware infection.

To Drain or Not to Drain

Draining the deep layers below the musculature should not be necessary as a muscle splitting or sparing technique should not allow for large potential spaces. Subcutaneous drains above the musculature should be considered in all patients with subcutaneous flaps with significant potential space. As this operation is typically performed after blunt trauma to the chest wall, the degree of soft tissue edema and contusion can weigh into the decision to drain the subcutaneous space. Drains above the muscle have not been shown to increase rates of hardware infection. They can, however, reduce the incidence postoperative fluid collections.

Adjuncts to SSRF

Bronchoscopy

Routine use of intraoperative bronchoscopy is controversial. We utilize bronchoscopy as an adjunct to pulmonary hygiene as this focused technique for secretion clearance is highly effective. Once intubated this allows for aggressive secretion clearance and culture sampling if clinically indicated. Due to pain and poor pulmonary mechanics preoperatively, most morbidity of rib fractures is from inadequate clearance of secretions. We utilize intraoperative intubation and mechanical ventilation to perform secretion clearance. SSRF greatly enhances a patient's ability to spontaneously clear secretions.

Thoracic Cavity Lavage

While limited thoracotomy and chest washout are not essential, we routinely perform this as an adjunct to SSRF. It eases fracture reduction and potentially eliminates the post-traumatic effusions and retained hemothorax common with rib fractures [4].

Analgesia

Additional analgesia should also be considered at the conclusion of SSRF. This can be achieved through a variety of techniques, including but not limited to cryoablation of the intercostal bundle, liposomal bupivacaine rib blocks, and insertion of indwelling infusion catheters. While each of these techniques has strengths and weaknesses, the general principle of striving for nonnarcotic pain control is the most important consideration [5].

References

1. Wikipedia. https://en.wikipedia.org/wiki/AO_Foundation
2. Pieracci FM, Coleman J, et al. A multicenter evaluation of the optimal timing of surgical stabilization of rib fractures. J Trauma Acute Care Surg. 2018;84(1):1–10.
3. Marasco. Analysis of bone healing in flail chest injury: do we need to fix both fractures per rib? J Trauma Acute Care Surg. 2014;77(3):452–8.
4. Majercik S, et al. Surgical stabilization of severe rib fractures decreases incidence of retained hemothorax and empyema. Am J Surg. 2015;210(6):1112–6. discussion 1116–7
5. Pieracci FM, et al. Consensus statement: Surgical stabilization of rib fractures rib fracture colloquium clinical practice guidelines. Injury. 2017;48(2):307–21.

13 Review of Currently Available Tools of the Trade

Silvana Marasco and Jose J. Diaz Jr.

The science of chest wall injury has continued to grow. The understanding of the mechanics of the chest wall and the various injury patterns has become better understood. This has allowed surgeons to study how to improve on the management of chest wall injury. The surge in technology on implant development had been significant in the last 10 years. In addition, the concept of absorbable prostheses for the purpose of rib fractures and sternal injuries has made it to the market.

Since the introduction of the laparoscope in the 1990s and the recognition of minimally invasive surgical approaches, these techniques have demonstrated improved patient outcomes. The latest advancements on chest wall injury have been related to minimally invasive approach. Surgeons are focusing on smaller incisions, sparing major muscle groups resulting in less damage, and a quicker return to function. As surgeons develop new surgical techniques and approaches, industry has had to innovate and make instruments to function in smaller sizes dementions.

Historical Repairs

There have been several procedures developed for external "stabilization of rib fractures", although these are now mostly of historical significance. Most of the time, the patient would have been placed in some type of skeletal traction or external K-wires left in place. The morbidity of the surgical procedure tended to be worse than the injury.

S. Marasco
Department of Cardiothoracic Surgery, The Alfred Hospital, Melbourne, VIC, Australia
e-mail: S.Marasco@alfred.org.au

J. J. Diaz Jr. (✉)
Division Acute Care Surgery, Department of Surgery, R Adams Cowley Shock Trauma, University of Maryland School of Medicine, Baltimore, MD, USA
e-mail: jdiaz@umm.edu

M. de Moya, J. Mayberry (eds.), *Rib Fracture Management*,
https://doi.org/10.1007/978-3-319-91644-6_13

The most common historical procedure for internal stabilization of rib fractures was the use of steel wires. The exposure was a standard thoracotomy incision and division of the latissimus dorsi and/or the serratus anterior muscle. There was only two-point stabilization which quickly failed. At some point, surgeons would attempt to wire several ribs together; this would create long-term restrictive disease problems. In the late 1990 and early 2000, surgeons began to use small fracture steel plates. These could technically be placed but where stiff and resulted in hardware pain.

In Europe, the development of the anterior clamping metal Judet plates was in use for many years. The challenges with these plates still required maximal dissection of the intercostal muscles and the periosteum. It is recognized that this results in potential bone ischemia and poor bone healing. In addition, the extensive exposure placed the neurovascular bundle at risk of injury (Fig. 13.1). In the USA, the stainless steel plates (Fig. 13.2) commonly used in small hand or forearm fractures were used for rib plating in the 1990s and early 2000s. The problem with the stainless steel plates was that they tended to be very rigid and over time proved to be more brittle. As a result, they tended to fail more often.

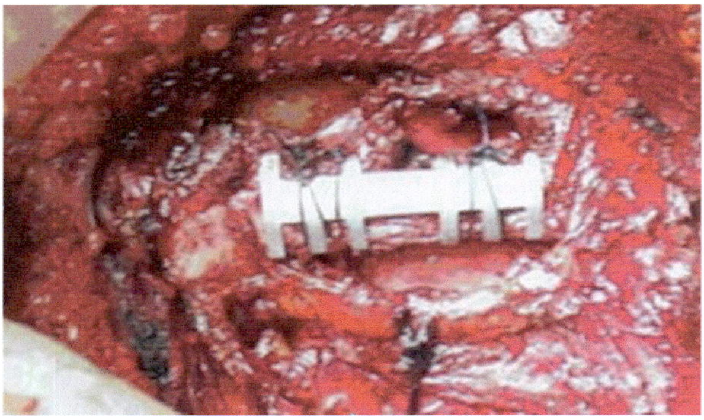

Fig. 13.1 With Judet plates the extensive exposure placed the neurovascular bundle at risk of injury

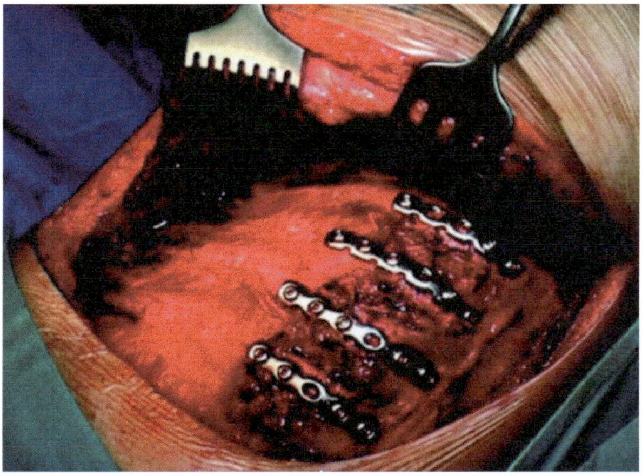

Fig. 13.2 Stainless steel plates

Absorbable Prostheses

The use of bioresorbable prostheses in surgical applications is not a new concept. Commonly used absorbable suture materials have been available for decades. Dexon (polyglycolide), Vicryl (polyglycolide/polylactide), PDS (polydioxanone) and Maxon (polyglycolide/trimethylene carbonate) are all absorbable suture materials used today. Over the past 20 years, these materials have expanded to new applications including small bone fixation with plates, screws, mesh and pins, drug delivery devices and even intravascular stents. The earliest biodegradable implants were typically made of a single polymer which led to problems of too rapid degradation (leading to early loss of strength in the implant and soft tissue reactions) or too slow degradation thus negating some of the benefits of the material [1, 2]. Current generation biodegradable devices are polymers, that is, several monomers are mixed to generate the optimal level of strength, malleability and degradation profile. The ratios of these monomers can be manipulated to adjust shear strength and torque resistance (e.g. in screws) and give optimal flexibility and tensile strength in products such as plates and mesh. The monomers currently used are L-lactide, D-lactide, glycolide and trimethylene carbonate.

Degradation of the polymer occurs initially by hydrolysis, with fragments being degraded to natural monomeric acids found in the body, such as lactic acid. These acids then enter the Krebs cycle and are metabolized into carbon dioxide and water. The different monomers degrade at different rates varying from several months to several years.

The use of bioresorbable bone fixation prostheses has a number of potential benefits over more rigid metal prostheses. Animal models using these absorbable plates have shown faster and stronger healing with absorbable plates compared to traditional metal plates which may actually slow bone healing [3]. This occurs because the metal plate protects the bone from any load, but in doing so it removes the stimulus for new bone growth (known as stress shielding). In contrast, the absorbable plates allow gradual transfer of stress loads to the bone, stimulating faster bone growth. Absorbable plates are also more flexible and so allow micro-movement at the fracture site which stimulates bone remodelling.

As the plates and screws are absorbable, there is no need to remove plates due to complications such as breakage which confer a potential cost saving to this procedure. Also, it minimizes the sequelae of possible prosthesis-related complications such as migration, palpability especially in thin patients or thermal sensitivity. Whereas metal prostheses usually mandate removal if these complications occur, with absorbable prostheses a more conservative approach can be adopted. Further, there is no contraindication to future magnetic resonance imaging in patients who have had these plates implanted in contrast to metal plates.

Polylactide copolymer prostheses have already been extensively used in a variety of low load-bearing craniofacial and orthopaedic applications. Currently there are no biodegradable prostheses specifically designed for rib fixation; however "off-label" use of biodegradable prostheses designed for other small bones has been used in this application.

A small pilot study and follow-up randomized controlled trial (RCT) used bioresorbable plates and screws for rib fixation in patients with flail chest [4, 5]. The prostheses used were Inion OTPS™ Mesh (Tampere, Finland), six-hole plates and screws designed for the fibula. The plates are malleable in hot water and fix in position as they cool. They maintain at least 40% of their strength at 3 months, by which time the fractures would be expected to have completely healed and are completely absorbed without toxicity over 1–3 years.

The RCT demonstrated superior clinical outcomes in patients whose ribs were fixed with the absorbable plates compared to medical management.

Use of the Inion resorbable mesh (Inion OTPS™, Tampere, Finland) has been studied in 32 acute rib fractured patients followed up by postal survey [6]. The mesh is used to wrap the rib fracture circumferentially and then fixed in position with absorbable sutures. About 63% of patients responded at a mean follow-up time of 1039 days post-fixation and reported low levels of pain. However 60% of those patients did complain of chest wall stiffness, although overall satisfaction with the operation was high. The authors did comment that the use of the resorbable mesh was technically advantageous in highly comminuted fractures as it allowed wrapping the multiple fragments in position. The mesh also facilitates the use of bone graft and cement. However, the disadvantage of the mesh wrap technique is that the intercostals have to be stripped from the rib. This involves a subperiosteal dissection to safely move the neurovascular bundle away from the inferior border of the rib, thus devascularizing the bone and potentially damaging the neurovascular bundle.

One of the advantages of resorbable materials is that they can be cut to size. Another resorbable mesh (Super Fixsorb; Takiron Co., Ltd., Osaka, Japan) has been used with absorbable sutures to fix divided ribs at the completion of minimally invasive cardiac surgery [7]. The authors reported no hardware failures and good clinical outcomes.

In a similar application, resorbable intramedullary pins have been used (Super Fixsorb; Ethicon Inc., Somerville, NJ) to reoppose divided ribs at the time of posterolateral thoracotomy [8]. However at 1 year follow-up, one third of patients were noted to have displacement at the fracture site, although this did not seem to impact clinical outcomes.

Hardware failure is one of the concerns of bioresorbable rib fixation systems. The application of prostheses for rib fixation is problematic because of the repetitive movement and load bearing of the bones being fixed. It is not possible to immobilize the affected area as it would be routine management in most other bone fractures. Although the ribs do not carry a heavy load, they are affected by torque in multiple directions due to the layers of intercostal muscles inserting onto the ribs. The forces impacting on the ribs are therefore constant and in multiple directions. Any rib fixation strategy needs to take these factors into account. The most analogous operative fixation strategy is the surgical management of a fractured mandible. This is another bone which is not immobilized after operative fixation, and although not constantly moving, it can be subjected to fairly high forces. Very good results have been reported with absorbable plating systems in this application [9, 10].

The technique for using absorbable prostheses is quite specific and different to the use of metal implants. There is more of a learning curve as the resorbable products need to be handled carefully. Because the screws are made of resorbable polymer, they are not strong enough to be self-tapping screws. Thus the technique firstly involves the drilling of the screw hole through both the outer and inner cortex, using the predefined screw holes in the plate as a guide to position. A depth gauge is then used to measure the thickness of the rib and thus choose the appropriate length screw. The drill hole then needs to be tapped, using a size just a millimetre or two smaller than the chosen screw diameter. The screw is then screwed into the pre-tapped hole. Rough handling during tapping can enlarge the hole leaving the screw loose. The other site of potential failure of the screws is the screw head. If too much torsion is applied to the screw head in trying to force the screw into the hole, the screw head can be twisted off.

A careful analysis of clinical reports of resorbable rib prostheses does show a significant incidence of hardware failure. In a series of ten patients with ribs fixed with ten-hole plates and screws (Macropore Advanced Resorbable Solutions, San Diego, CA), two patients were noted to have partial loss of rib fracture alignment after 24 h postoperatively [11]. The authors recommended that absorbable cerclage sutures were required to hold the plate in position in addition to screws.

In another study of 13 patients who had 58 ribs fixed with resorbable plates and screws (Inion Resorbables, Inion OTPS™, Tampere, Finland), ten rib fixation failures were noted [12]. These were particularly prominent when used posteriorly indicating that the forces generated by the chest wall posteriorly are more likely to exceed the ultimate tensile strength of the absorbable prostheses used in that study.

Thus the use of absorbable prostheses has many potential advantageous qualities in rib fixation applications. However, the current reported rate of hardware failure seems unacceptably high. It should be kept in mind though that none of these published reports of rib fixation with absorbable prostheses have used products designed specifically for rib fixation. It should be possible to increase strength in these prostheses while maintaining appropriate flexibility by altering the monomer components. This is an area currently under development.

RibLoc®: Rib Fracture Plating System and U Plus Chest Wall Plating System

The titanium RibLoc U plate (Acute Innovations, LLC, Hillsboro, OR) is a short 4.5 cm plate designed to be placed on the rib via less invasive incisions. The superior aspect of the rib needs to be cleared of muscle attachments to place the plate over the rib, but the inferior aspect of the rib remains untouched, thus protecting the neurovascular bundle. The plate is then locked to the rib with bicortical screws which also lock into the plate posteriorly, thus providing an extremely strong construct. The plate is designed to only require four screws for fixation. Theoretically the construct can be used on osteoporotic bone as the screws lock into the plate on both sides of the rib. The challenge is the amount of force/load that must be distributed over the plate from the continuous breathing or motion of the patient that is unknown (Figs. 13.3 and 13.4).

In a benchtop study of human cadaveric ribs, the U plate construct was noted to exhibit superior stiffness to a titanium anterior plating construct alone when cycled 50,000 times at a load of ±2 N at 1 Hz to simulate 48 h of deep breathing [13].

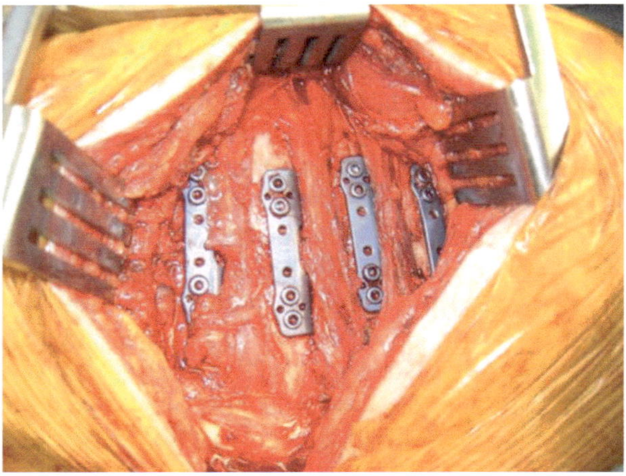

Fig. 13.3 RibLoc system (Acute Innovations, LLC, Hillsboro, OR)

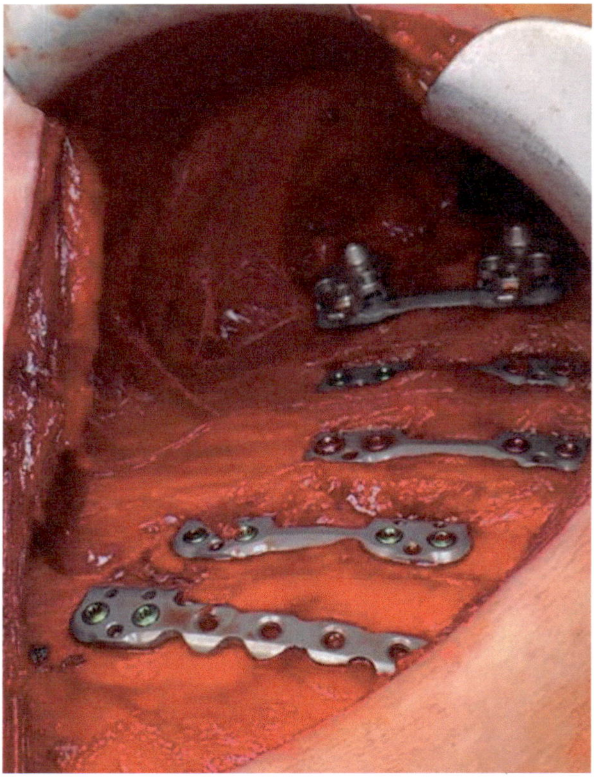

Fig. 13.4 RibLoc U plus—compressible U plate

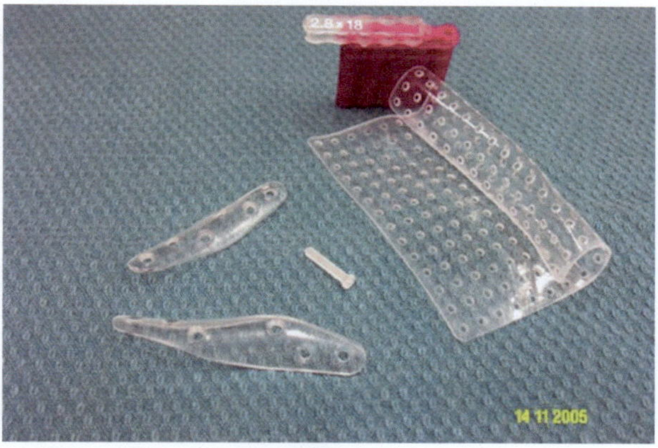

Fig. 13.5 Inion resorbable prostheses used. Mesh, plate and screws are pictured. The mesh is not designed for strength and was only used in combination with the plates (Inion, Finland)

Since the initial design, the U plate has been modified to improve stability and usability. The screws are now dual locking screws that lock into both the anterior and posterior plate (Fig. 13.5). The initial intention of the plate was always that it could be bent in the midsection to follow the curvature of the rib. The thickness of the rib is then measured and a plate with the corresponding depth of the U section chosen to fit over the rib (available in 6, 8, 10, 12 and 14 mm). However the stiffness of the plate and in particular the U section made it difficult to achieve a perfect fit over sections of rib with a tighter radius of curvature, even with bending. Longer plates were then marketed to allow bridging of comminuted fractures. The currently available RibLoc U Plus system now has a compressible U section so that the length of screw (chosen according to rib thickness) determines the anterior-posterior distance on the U plate. Further bicortical screws are placed along the expanded plates to fixate any bone fragments.

The changes with the current version of the U plate which are the degree of stiffness are less. This allows the plate to be more flexible. The degrees of rotation do require manual twisting.

Anterior Cortical Plates

MatrixRIB™ Fixation System

The MatrixRIB™ titanium fixation system (DePuy Synthes, Comp, West Chester, PA) consists of precontoured plates designed to sit on the outer cortex of the ribs (Fig. 13.6). The curvature of the ribs is quite complex with an in-plane and out-plane curvature as well as twist along the length of the rib [14]. The radius of curvature also changes for each rib, increasing from 1 to 12 as well as increasing along the length of the rib from posterior to anterior. This helps explain why previously used generic small metal plates have had such poor results. Without contouring the plate to the complex curvature of the rib, the risk of hardware lift-off and screw pullout is very high. The MatrixRIB™ plates can be further bent with specific bending tools to ensure

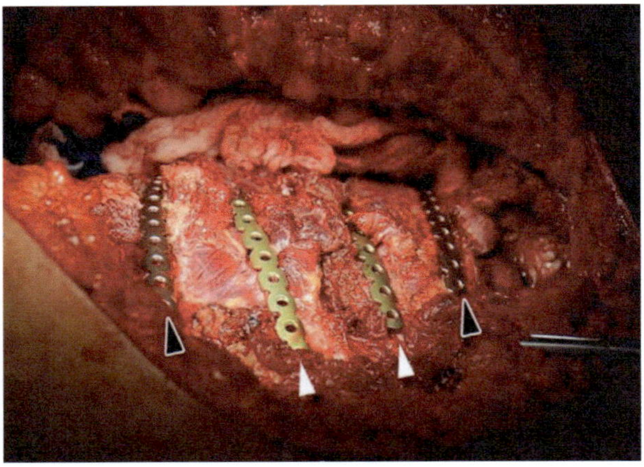

Fig. 13.6 MatrixRIB™ titanium fixation system (DePuy Synthes, Comp, West Chester, PA)

perfect apposition of the plate to the bone. The titanium screws are designed for bicortical fixation as well as locking to the plate, with a recommendation that three screws on either side of the fracture be used. The plates are quite long and should be cut to length necessary to overlap a fracture and the supporting square placement.

Most of the plates are designed to cover the majority of a rib. As a result, placing these plates requires the ability to decide how much overlap of a plate is needed on each side of a fracture. Also, if one plans to cover a short flail segment, one may need to use the entire plate. This will make the rib stiffer and over time can contribute to hardware pain.

RibFix Blu™ Thoracic Fixation System

The Zimmer Biomet RibFix Blu™ is similar to the Synthes MatrixRIB system. RibFix Blu™ is a comprehensive thoracic fixation system that allows for the stabilization and rigid fixation of fractures in the chest wall including sternal reconstructive procedures, trauma or planned osteotomies. The system may be used in normal and poor bone to promote bone union. They are thinner and lighter which in theory would make them more flexible. They do require rotational twisting to match the rib, and the system does have self-tapping screws. The time benefit of not needing to drill the holes does result in less operative time.

These are anterior plates made of titanium. These are similar to the Synthes, MatrixRIB and the Zimmer Biomet as they are considered "anterior plates". These are positioned on the external surface of the rib. The plates must be cut and contoured to the rib. As with all anterior plates, these require three screws on each side of the fracture.

StraCos

StraCos (Strasbourg Thoracic Osteosyntheses System—STRATOS™; MedXpert GmbH, Heitersheim, Germany) clips and connecting bars have been designed for chest wall reconstruction either following traumatic fractures or for chest wall reconstruction after surgical resection. The pure titanium prostheses are designed to be crimped over the superior and inferior cortex of the rib using a specialized crimping tool. The 3D rib clip (pink prosthesis) is used to span simple fractures and comes in a standard and XL width with 6, 9 and 13 segment variants. The gold prostheses (rib clip with connecting part) are used to span defects and multiple fractures. The gold prostheses come as two clips with either a straight or angulated connecting segment which then connects to a serrated spanning bar which can be cut to the appropriate length. The angulated segments are available as either a 45° or 22.5° angle.

In a case-matched controlled study of ten polytrauma patients undergoing rib stabilization with StraCos prostheses, ventilator time and overall hospital stay were significantly reduced compared to the control patients [15]. A median of three [2–6] plates were used per patient with an operative philosophy of fixing one out of every two fractures. In an application using the extended rib clips with connecting bars, a recent case report was able to demonstrate preserved bucket handle motion of the ribs in a patient who had had an extensive chest wall resection for sarcoma [16]. A large case series of 94 trauma patients treated exclusively with the StraCos rib fixation system has been reported [17]. A median of four ribs was stabilized per patient, with no hardware failures reported during the 6-month follow-up. Two patients had clips removed for infection. The authors focused their postoperative follow-up on respiratory mechanics and potential for restriction on the operated side. Although 19% of patients who were operated on/for flail chest reported stiffness on the operated side, dynamic magnetic resonance imaging analysis of vital capacity on the operated side versus a target value derived from the nonoperated side at 6 months showed a group mean of 92%. The authors interpreted this to indicate good preservation of chest wall movement, although there was only a modest correlation with forced vital capacity. Thirteen patients had ongoing neurological pain on the side of the surgery.

KLS Martin Rib Fixation

KLS Martin (Jacksonville, FL) has developed a titanium "convergent biaxial 3D fixation" system based on contoured plates and self-tapping unicortical screws. The aim of this system is to reduce the workload for the surgeon by using self-tapping screws, without losing strength or stability that is a risk with using only unicortical screws. By alternating the angle that each screw enters the bone, the risk of hardware pullout is theoretically reduced maintaining fixation of the fracture.

Intramedullary Fixation

The idea of intramedullary fixation of fractured ribs is not completely new. Kirschner wires have been described in rib fixation for many years. However, Kirschner wires were designed for small bones which would then be immobilized (e.g. with a plaster of Paris cast) to allow healing. The use of Kirschner wires in rib fixation has led to multiple reports of wire migration and cortical cut-out. They also provide poor rotational and longitudinal stability which is particularly important in an application where the bone cannot be completely immobilized in the postoperative setting.

A commercially available titanium intramedullary splint has been designed specifically for rib fixation (DePuy Synthes, West Chester, PA, USA). The prosthesis is a titanium rib splint 97 mm long with a rectangular cross section of 1 mm thick and 3, 4 or 5 mm wide. The rectangular cross section reduces rotational instability, and the titanium construct maintains some flex so that it can follow the curvature of the rib as it is reamed through the intramedullary canal. The device is inserted into the medulla of the rib via an opening in the outer cortex that is drilled approximately 30 mm away from the fracture site. The device is advanced along the intramedullary canal, traversing the rib fracture with its tapered leading edge designed to reduce insertion forces. Although the initial part of the rib splint is straight, the remainder has a curvature of 200-mm radius designed to approximate the curvature of the ribs and to minimize residual stress after insertion. The trailing end of the splint remains on the external aspect of the outer cortex and has a threaded screw hole for fixation of the splint to the rib with a single bicortical locking screw. There is no distal fixation as such the stabilization of the rib fracture being achieved by the stiffness of the splint within the intramedullary canal (Fig. 13.7).

Finite element analysis has shown that there is micro-movement at the fracture site which is splinted with an intramedullary prosthesis without distal fixation [18]. This is thought to be beneficial as micro-movements at fracture sites are known to stimulate osteoblast formation and promote bone healing [19, 20].

In a benchtop study, cadaveric ribs were loaded to failure to induce realistic rib fractures [21]. The load was applied to the rib specimen to recreate an anatomical anteroposterior force. The rib fractures were fixed with the titanium intramedullary rib splint, and the fixation constructs were dynamically loaded to 360,000 cycles with an exaggerated respiratory loading magnitude of 200 N/mm (representing five times the bending moment measured in vivo on human ribs during physiological respiration). Constructs failed by splint bending in 44% of specimens and by developing fracture lines along the superior and inferior cortices in 56% of specimens. However, regardless of the failure mode, all rib splint constructs recoiled elastically after failure and retained functional reduction and fixation. No construct exhibited implant cut-out or migration through the lateral cortex.

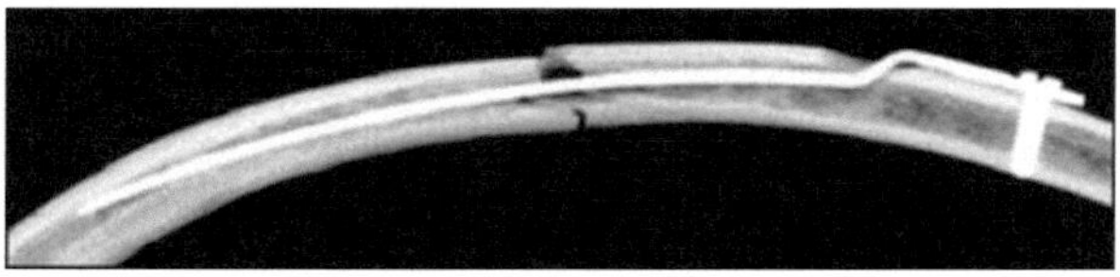

Fig. 13.7 The stabilization of the rib fracture is achieved by the stiffness of the splint within the intramedullary canal

In a further study by the same authors, the performance of the intramedullary splint was compared to Kirschner wires in 22 paired human ribs [22]. Tested in a similar benchtop setup to the above study, the Kirschner wire constructs were found to exhibit three times more subsidence than the titanium rib splint constructs. All Kirschner wire constructs failed during testing either by wire migration or bending, some with complete loss of stability of the construct. In contrast, the titanium intramedullary splints failed by development of fracture lines along the superior and inferior cortices, but all the constructs maintained functional reduction and fixation.

The intramedullary rib splint, although recommended as an option for single rib fractures, does lend itself to fixation of more difficult to access rib fractures, particularly those under the scapula where outer cortical hardware would be problematic due to the limitations of space. In a clinical study of 33 intramedullary splints inserted in 14 patients with multiple rib fractures, non-healing was noted in only two of the fractures (6%) [18]. In both patients, there was a failure at the rib/splint interface with removal of the hardware prior to 3 months, residual deformity and failure of union. No other patient was residual deformity- or nonunion noted at a fracture site fixed with an intramedullary splint. There were no hardware failures in this cohort. By 3 months post-surgery, 94% of patients had at least partial healing with good alignment and adequate fracture stabilization. By 6 months the fractures, which had shown partial healing, had all completely healed. There were no late failures (between 3 and 6 months) of either hardware or rib/splint interfaces.

Another case report has identified posterior migration of a rib splint through the lateral cortex of the rib leading to disabling pain, resolved only by removal of the splint [23]. The use of titanium elastic nails in an off-label application for intramedullary rib fixation has also been described with good results [24].

For the splints to work best, they must be well matched to the individual patient's rib thickness and the not be osteoporotic bones. If the splints are not appropriately matched, they can result in too much movement and further injury of the bone.

Lastly, intermediary splints work only in the setting where the local forces acting on a rib fracture are keeping the ends of the rib together. There are three types of injuries and distractions. There is outward distraction that results in the ends of the fracture being pulled apart. There is inward distraction with the forces pulling the ends of the rib fracture together.

Conclusion

The current market has several options for stabilizing rib fractures. None of the implants have been specifically studied for "best" indications based on injury type, bone density, cortical thickness or long-term hardware pain. We hope to see more research in the future as the focus on this injury has developed. The future directions are underway for thoracoscopic approaches and internal rib plating for stabilization.

References

1. Andriano KP, Pohjonen T, Tormala P. Processing and characterization of absorbable polylactide polymers for use in surgical implants. J Appl Biomater. 1994;5(2):133–40.
2. Pietrzak WS, Sarver DR, Verstynen ML. Bioabsorbable polymer science for the practicing surgeon. J Craniofac Surg. 1997;8(2):87–91.
3. Viljanen J, Pihlajamäki H, Kinnunen J, Bondestam S, Rokkanen P. Comparison of absorbable poly-L-lactide and metallic intramedullary rods in the fixation of femoral shaft osteotomies: an experimental study in rabbits. J Orthop Sci. 2001;6:160–6.
4. Marasco S, Cooper J, Pick A, Kossmann T. Pilot study of operative fixation of fractured ribs in patients with flail chest. ANZ J Surg. 2009;79(11):804–8.
5. Marasco SF, Davies AR, Cooper J, Varma D, Bennet V, Nevill R, Lee G, Bailey M, Fitzgerald M. Prospective randomized controlled trial of operative rib fixation in traumatic flail chest. J Am Coll Surgeons. 2013;216(5):924–32.
6. Campbell N, Conaglen P, Martin K, Antippa P. Surgical stabilization of rib fractures using Inion OTPS wraps-techniques and quality of life follow up. J Trauma. 2009;67(3):596–601.
7. Ito T, Kudo M, Yozu R. Usefulness of osteosynthesis device made of hydroxyapatite-poly-L-lactide composites in post-access cardiac surgery. Ann Thorac Surg. 2008;86:1905–8.
8. Kawachi R, Watanabe S, Suzuki K, Asamura H. Clinical application of costal coaptation pins made of hydroxyapatite and poly-L-lactide composite for posterolateral thoracotomy. Eur J Cardiothorac Surg. 2008;34:510–3.
9. Bell RB, Kindsfater CS. The use of biodegradable plates and screws to stabilize facial fractures. J Oral Maxillofac Surg. 2006;64:31–9.
10. Yerit KC, Hainich S, Turhani D, Klug C, Wittwer G, Ockher M, Ploder O, Undt G, Baumann A, Ewers R. Stability of biodegradable implants in treatment of mandibular fractures. Plast Reconstr Surg. 2005;115:1863–70.
11. Mayberry JC, Terhes JT, Ellis TJ, Wanek S, Mullins RJ. Absorbable plates for rib fracture repair: preliminary experience. J Trauma. 2003 Nov;55(5):835–9.
12. Marasco SF, Sutalo ID, Bui AV. Mode of failure of rib fixation with absorbable plates: a clinical and numerical modelling study. J Trauma. 2010;68(5):1225–33.
13. Sales JR, Ellis TJ, Gillard J, Liu Q, Chen JC, Ham B, Mayberry JC. Biomechanical testing of a novel, minimally invasive rib fracture plating system. J Trauma. 2008;64(5):1270–4.
14. Mohr M, Abrams E, Engel C, Long WB, Bottlang M. Geometry of human ribs pertinent to orthopaedic chest-wall reconstruction. J Biomech. 2007;40:1310–7.
15. Jayle CP, Allain G, Ingrand P, Laksiri L, Bonnin E, Hajj-Chahine J, Mimoz O, Corbi P. Flail chest in polytraumatized patients: surgical fixation using stracos reduces ventilator time and hospital stay. Biomed Res Int. 2015;2015:624723.
16. Coonar AS, Wihlm JM, Wells FC, Qureshi N. Intermediate outcome and dynamic computerised tomography after chest wall reconstruction with the STRATOS titanium rib bridge system: video demonstration of preserved bucket-handle rib motion. Interact Cardiovasc Thorac Surg. 2011;12(1):80–1.
17. Wiese MN, Kawel-Boehm N, Moreno de la Santa P, Al-Shahrabani F, Toffel M, Rosenthal R, Schäfer J, Tamm M, Bremerich J, Lardinois D. Functional results after chest wall stabilization with a new screwless fixation device. Eur J Cardiothorac Surg. 2015;47:868–75.
18. Marasco S, Quayle M, Summerhayes R, Sutalo ID, Liovic P. An assessment of outcomes with intramedullary fixation of fractured ribs. J Cardiothorac Surg. 2016;11(1):126–40.
19. Claes L, Heigele CA. Magnitudes of local stress and strain along bony surfaces predict the course and type of fracture healing. J Biomech. 1999;32:255–66.
20. Lacroix D, Predergast P. A mechano-regulation model for tissue differentiation during fracture healing: analysis of gap size and loading. J Biomech. 2002;35:1163–71.
21. Helzel I, Long W, Fitzpatrick D, Madey S, Bottlang M. Evaluation of intramedullary rib splints for less-invasive stabilisation of rib fractures. Injury. 2009;40:1104–10.

22. Bottlang M, Helzel I, Long W, Fitzpatrick D, Madey S. Less-invasive stabilization of rib fractures by intramedullary fixation: a biomechanical evaluation. J Trauma. 2010;68(5):1218–24.
23. Zaidenberg EE, Rossi LA, Bongiovanni SL, Tanoira I, Maignon G, Ranalletta M. Snapping scapular syndrome secondary to rib intramedullary fixation device. Int J Surg Case Rep. 2015;17:158–60.
24. Tarng Y-W, Liu Y-Y, Huang F-D, Lin H-L, Wu T-C, Chou Y-P. The surgical stabilization of multiple rib fractures using titanium elastic nail in blunt chest trauma with acute respiratory failure. Surg Endosc. 2016;30:388–39.

Postoperative Complications After Rib Fracture Repair

14

Raminder Nirula

Complications following rib fracture fixation are difficult to distinguish from the complications related to the natural history of the underlying injury. Patients requiring rib fracture fixation are at risk for pulmonary-related complications, but teasing out, which of these complications is directly the result of the rib fracture repair itself, is difficult. Additionally, these patients are at risk for extrathoracic complications associated with their injury severity such as deep venous thrombosis, pulmonary embolus, catheter-related blood stream infection, urinary tract infection, and injury-specific complications.

Since rib fractures are frequently associated with other injuries that increase the risk of death and complications, an appropriate risk adjustment is necessary. In order to determine whether rib fracture fixation is associated with morbidity or mortality, one must first examine the rates in those managed nonoperatively.

Complications that may be related to the chest trauma in general or to the surgical fixation procedure itself include pneumonia, prolonged ventilation, prolonged hospitalization, and death [1–9]. In patients where flail chest was the principal diagnosis, in-hospital mortality was 4.2% for those 45–64 years of age, 11.3% for those 65–84, and 28.4% for those 85 or older. A number of studies have confirmed the increased mortality risk of rib fractures in the elderly when compared to younger patients [10–13]. A meta-analysis of 29 studies of blunt thoracic trauma identified that age at ≥65 years, three or more rib fractures, the presence of preexisting disease, and the development of pneumonia post-injury were significant risk factors for mortality [14]. Critical care resources are required by as many as 40% of multiple rib fracture patients, the majority of whom are elderly (68%) [15]. The mean number of mechanical ventilation days required for patients with multiply displaced rib

R. Nirula
Acute Care Surgery Section, Department of Surgery, University of Utah, Salt Lake City, UT, USA
e-mail: r.nirula@hsc.utah.edu

M. de Moya, J. Mayberry (eds.), *Rib Fracture Management*,
https://doi.org/10.1007/978-3-319-91644-6_14

fractures or flail chest ranges from 7 to 30 days [2, 16–18]. Nosocomial pneumonia occurs significantly more frequently in elderly patients than in their younger counterparts with a dose-response relationship showing higher rates of nosocomial pneumonia in those with greater numbers of rib fractures (Fig. 14.1 and Table 14.1) [1]. While pulmonary contusion contributes to the length of mechanical ventilation, the severity of rib fractures independently predicts the duration of mechanical ventilation. These morbidity and mortality rates must be considered when measuring such rates in those who undergo rib fracture fixation to determine the fixation-specific frequency of these events.

It appears that rib fracture fixation does not increase mortality and potentially may be associated with improved survival. A meta-analysis pooling the results of nine studies that included over 500 patients showed a 56% lower risk of death for those who underwent rib fracture fixation. This same analysis showed a 55% lower risk of pneumonia and 75% lower likelihood of tracheostomy [4]. It is important to note that this meta-analysis included both randomized controlled trials as well as observational studies which carry some risk of inadequate risk adjustment. Based upon the current data, however, it does not appear that rib fracture fixation increases the risk of mortality, pneumonia, or need for tracheostomy.

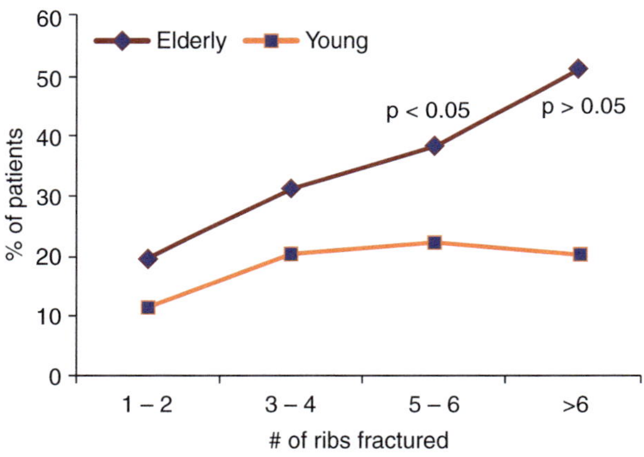

Fig. 14.1 Pneumonia by number of rib fractures in elderly and younger patients (from Bulger EM, et al. Rib fractures in the elderly. J Trauma. 2000;48(6):1040–6; discussion 1046–7, with permission)

Table 14.1 Common complications attributable to surgical fixation

Complication	Rate, %
Wound infection	0–10
Hardware infection	0–2.9
Empyema	5
Chest tightness/dyspnea	19–33
Chronic pain	11
Hardware migration	0–2.9

In fact, postoperative complications specific to the procedure of rib fracture fixation are rare. These complications can be divided into early postoperative and late complications. Common early complications include wound infection, empyema, hematoma, or persistent effusion [19]. Late complications include chest tightness, dyspnea, chronic pain, hardware dislodgement, hardware infection, and periprosthetic fracture. It is often difficult to determine whether chest tightness, dyspnea, or chronic pain occurs as a result of fixation or underlying chest wall trauma.

It is difficult to discern whether or not the observed complications are related to the surgical fixation or the chest injury itself. In 2012, a decision analysis was performed to determine the cost-effectiveness of rib fracture fixation for flail chest which included the commonly reported complication rates associated with rib fracture fixation [5]. In that review an assessment of the National Trauma Databank demonstrated that among surgical fixation patients, prolonged intubation (>96 h) occurred at a rate of 17% and the rate of VAP in tracheostomy patients was 75% versus 25% in those without a tracheostomy [20]. These rates were less than or comparable to the rates observed in nonoperative patients suggesting that fixation is not associated with increased risk for prolonged intubation or VAP. In a study of 66 anterolateral flail chest patients undergoing surgical fixation from 1990 to 1999, five patients (7.6%) developed pneumonia postoperatively. Four patients died of ARDS (6%) and multi-organ failure. Two patients (3%) developed wound infections but did not require plate removal. At 6-month follow-up, attenuated sensitivity of the anterior chest wall was present in 7% and 11% of patients who suffered from persistent pain at the operative site. Half of these patients had their plates removed which were associated with improvement in their pain. The entire cohort had radiologic follow-up at 6 months which did not show any plate dislocation; however asymptomatic dislocation of screws was seen in two patients (3%). This study also provided pulmonary function test results at 6 months showing that 22% had an obstructive pattern and 16% had a mixed obstruction and restriction pattern [21]. Unfortunately this study did not report on these outcomes for flail chest patients who did not undergo fixation to determine if surgical fixation improved or worsened these outcomes. Granetzny et al. evaluated pulmonary function in a randomized trial of 40 flail chest patients in whom half underwent surgical fixation with wires and half were treated nonoperatively. In 2 months post-injury, the surgical group had a higher forced vital capacity and total lung capacity indicating a less restrictive pattern than those who did not undergo fixation [3]. Likewise, a study by Ohresser et al. reported a lower incidence of dyspnea in surgical patients compared to nonoperative patients at 1-year follow-up [22]. In Tanaka's randomized study, chest tightness at 1 year was 2.5 times more common in nonoperative patients compared to operative patients [16]. Chest tightness was reported at 19% but with no evidence of a restrictive breathing pattern on pulmonary function tests in a series of flail chest patients undergoing rib fracture fixation at 6-month follow-up [23]. These results indicate that rib fracture fixation does not reduce lung expansion either objectively or subjectively compared to nonoperative management in the flail chest population. Whether or not this holds true for patients who undergo fixation for significant fracture displacement in the absence of flail chest has yet to be determined.

Infectious complications are of significant concern when placing hardware as this may require surgical removal or prolonged treatment courses. In Granetzny's randomized trial, there was a significantly lower chest infection rate of 10% in the surgical group compared to 50% in the nonoperative group. Empyema occurred in 5% of the surgical group and 10% of the nonoperative group. There were two cases (10%) of mediastinitis in the surgical group and two cases (10%) of wound infection. None of the operative patients had a pulmonary embolism, and there was one pulmonary embolism in the nonoperative group. The authors did not report on whether there were any issues with wire migration; however the follow-up period was relatively short. [3]

Complications that can be attributed to the surgical procedure itself center around wound infection, hardware migration, and hardware infection. In a matched case-control series comparing 22 patients undergoing fixation with a locking plate mechanism to 28 nonoperative patients, followed for an average of 17 months, there were no cases of hardware failure, hardware prominence, wound infection, or nonunion [24]. In a randomized trial of 23 rib fracture fixation compared to 23 nonoperative management from Australia, there was no comment on plate migration or failure or wound complications. This study had one patient in each group with concerns about the cosmetic appearance of their chest, and there was no difference in pulmonary function tests or 3- and 6-month quality of life scores. In a review of 68 flail chest patients undergoing operative fixation with a traditional wide thoracotomy approach using Stratos titanium rib clips with interlocking stabilization rods, wound infection rate was 2.9%, and two patients required partial hardware removal due to infection. All patients had radiologic assessment at 6 months which showed no evidence of fixation failure or plate migration [23]. Pieracci et al. performed a review of 35 patients undergoing fixation for flail chest or significantly displaced rib fractures in which there was one case (2.9%) of hardware infection requiring removal and one case of a radiographically detected screw dislodgement that was incidentally identified. Since this study only evaluated patients to hospital discharge, we do not know if further fixation failures might have occurred in the long term.

Concerns regarding the use of Kirschner wires for rib fracture repair revolve around their possible migration and cutting through the rib during repeated movement with breathing. In a biomechanical study, intramedullary splinting was compared to Kirschner wire fixation using 22 paired human ribs undergoing repeated loading after fracture fixation. There was three times more subsidence after loading in the wire fixation ribs compared to the splint fixation ribs. Additionally, splints remained 48% stronger, and almost half of the wire ribs suffered catastrophic failure with the wire cutting through the medial cortex. While there were no catastrophic failures in the splint group, these ribs did show evidence of development of fracture lines along the superior and inferior cortex. [25] In a porcine rib fracture fixation biomechanical comparison of intramedullary screws to the Inion orthopedic trauma plating system, the plating system maintained fixation at higher breaking forces than the intramedullary screw [26]. Based upon biomechanical data and limited clinical data, therefore, the lowest rate of fixation failure may be associated with plating systems as opposed to intramedullary screws or wires.

References

1. Bulger EM, et al. Rib fractures in the elderly. J Trauma. 2000;48(6):1040–6. discussion 1046–7
2. Balci AE, et al. Open fixation in flail chest: review of 64 patients. Asian Cardiovasc Thorac Ann. 2004;12(1):11–5.
3. Granetzny A, et al. Surgical versus conservative treatment of flail chest. Evaluation of the pulmonary status. Interact Cardiovasc Thorac Surg. 2005;4(6):583–7.
4. Leinicke JA, et al. Operative management of rib fractures in the setting of flail chest: a systematic review and meta-analysis. Ann Surg. 2013;258(6):914–21.
5. Bhatnagar A, Mayberry J, Nirula R. Rib fracture fixation for flail chest: what is the benefit? J Am Coll Surg. 2012;215(2):201–5.
6. Nirula R, et al. Rib fracture stabilization in patients sustaining blunt chest injury. Am Surg. 2006;72(4):307–9.
7. Marasco SF, et al. Prospective randomized controlled trial of operative rib fixation in traumatic flail chest. J Am Coll Surg. 2013;216(5):924–32.
8. Majercik S, et al. Long-term patient outcomes after surgical stabilization of rib fractures. Am J Surg. 2014;208(1):88–92.
9. Slobogean GP, et al. Surgical fixation vs nonoperative management of flail chest: a meta-analysis. J Am Coll Surg. 2013;216(2):302–11. e1
10. Bauza G, et al. High mortality in elderly drivers is associated with distinct injury patterns: analysis of 187,869 injured drivers. J Trauma. 2008;64(2):304–10.
11. Brasel KJ, et al. Rib fractures: relationship with pneumonia and mortality. Crit Care Med. 2006;34(6):1642–6.
12. Kent R, Woods W, Bostrom O. Fatality risk and the presence of rib fractures. Ann Adv Automotive Med. 2008;52:73–82.
13. Sharma OP, et al. Perils of rib fractures. Am Surg. 2008;74(4):310–4.
14. Battle CE, Hutchings H, Evans PA. Risk factors that predict mortality in patients with blunt chest wall trauma: a systematic review and meta-analysis. Injury. 2012;43(1):8–17.
15. Sirmali M, et al. A comprehensive analysis of traumatic rib fractures: morbidity, mortality and management. Eur J Cardio-Thoracic Surg. 2003;24(1):133–8.
16. Tanaka H, et al. Surgical stabilization of internal pneumatic stabilization? A prospective randomized study of management of severe flail chest patients. J Trauma. 2002;52(4):727–32. discussion 732
17. Voggenreiter G, et al. Operative chest wall stabilization in flail chest—outcomes of patients with or without pulmonary contusion. J Am Coll Surg. 1998;187(2):130–8.
18. Ahmed Z, Mohyuddin Z. Management of flail chest injury: internal fixation versus endotracheal intubation and ventilation. J Thorac Cardiovasc Surg. 1995;110(6):1676–80.
19. Nirula R, Mayberry JC. Rib fracture fixation: controversies and technical challenges. Am Surg. 2010;76(8):793–802.
20. National Trauma Data Bank. American College of Surgeons.
21. Lardinois D, et al. Pulmonary function testing after operative stabilisation of the chest wall for flail chest. Eur J Cardiothorac Surg. 2001;20(3):496–501.
22. Ohresser P, et al. The functional sequelae of closed thoracic injuries (apropos of 92 cases). Poumon Coeur. 1972;28(3):145–50.
23. Wiese MN, et al. Functional results after chest wall stabilization with a new screwless fixation device. Eur J Cardio-Thoracic Surg. 2015;47(5):868–75.
24. Althausen PL, et al. Early surgical stabilization of flail chest with locked plate fixation. J Orthop Trauma. 2011;25(11):641–7.
25. Bottlang M, et al. Less-invasive stabilization of rib fractures by intramedullary fixation: a biomechanical evaluation. J Trauma. 2010;68(5):1218–24.
26. Campbell N, Richardson M, Antippa P. Biomechanical testing of two devices for internal fixation of fractured ribs. J Trauma. 2010;68(5):1234–8.

15 Billing and Coding

Andrew Doben

Medical billing and coding have become increasingly complex in the current state of health care. Although clinicians are responsible for determining the procedures, diagnoses, and evaluation and management codes for patient interactions, this does not tend to be their primary focus. As such, when instituting new procedures and operations, it is highly recommended that clinicians partner with specialists in billing and coding. Various certifications exist for this highly specialized group of trained individuals. Depending on your practice setting, these experts may be available to clinicians as either in-house resources or independent consultants. Ongoing, active dialogue between well-trained coders, compliance specialists, and surgeons will improve success with reimbursement submission.

General Principles

This chapter will focus on the coding related to the operative procedures of surgical stabilization of rib fractures and other procedures typically encountered while performing open reduction and internal fixation of ribs. As with all operative procedures, complete and detailed operative reports that become a permanent part of the patient's medical record will also provide the documentation necessary to support your billing and reimbursement. Every aspect of the surgical procedure must be described clearly and contain the necessary components that are described in the American Medical Association (AMA) and Current Procedural Terminology (CPT) code descriptions.

A. Doben
Division of Trauma, Acute Care Surgery and Surgical Critical Care, Department of Surgery, University of Massachusetts Medical School–Baystate, Springfield, MA, USA
e-mail: Andrew.doben@baystatehealth.org

M. de Moya, J. Mayberry (eds.), *Rib Fracture Management*,
https://doi.org/10.1007/978-3-319-91644-6_15

According to the Joint Commission standards, all operative procedures require an immediate postoperative note, as well as a more detailed operative report that must be dictated or handwritten immediately within 24 h following the procedure [1].

The immediate postoperative note must contain the following elements:

- Name of primary surgeon/physician and assistants
- Preoperative diagnosis
- Postoperative diagnosis
- Name of the procedure performed
- Findings of the procedure
- Specimens removed
- Estimated blood loss
- Date and time recorded

Documenting the time recorded is very important as it confirms that the note was completed prior to moving the patient to the next level of care.

As noted above, a more detailed operative note is also required and must provide the necessary documentation to ensure that the CPT code matches the procedure performed.

The full operative report must contain all of the elements of the immediate post-operative note as well as the following:

- Indications for the procedure
- Intraoperative complications
- A full description of the procedure

The level of detail in the full operative report will directly affect reimbursement; in general, the more detailed your report, the better your reimbursement. Each operative note should include detail about patient positioning, preparation, and draping. Description of time-out procedures, verification of patient and site, as well as administration of antimicrobial agents should be included. The conclusion of the note should always include information about the sponge, needle, and instrument counts and review of intraoperative complications not specific to the procedure (i.e., anesthesia issues). Patient tolerance of the procedure and current stability level, as well as planned disposition location (ICU versus recovery room), should also be included.

The authors recognize that it can be difficult to be certain which rib level is actually being reconstructed in the setting of the severely injured chest and this will become even more challenging as these procedures become more minimally invasive. As done in many centers, we perform postoperative chest radiographs on a routine basis, and we have become accustomed to completing our operative note dictation after reviewing this radiograph as it can assist in increasing the detail of fractures repaired, the status of the hemithorax, the position of any postsurgical tubes and drains, and final positioning of all inserted hardware.

Basic Coding Principles

The Centers for Medicare and Medicaid Services (CMS) developed the National Correct Coding Initiative (NCCI) to ensure proper coding for reimbursement. These NCCI edits have been adopted by most private carriers to ensure that proper payment occurs based on coding descriptions, particularly CPT coding as delineated in the AMA CPT manual. From a practical standpoint, this determines which procedures are "bundled" or included within the main procedure being performed. There are many available resources online to verify the NCCI based on the CPT codes [2].

Open Reduction and Internal Fixation Codes

In 2015, the AMA approved category 1 CPT codes for open reduction and internal fixation of rib fractures, and three codes were introduced to describe this procedure. These codes refer to a unilateral procedure, and bilateral repairs are coded with a modifier.

21811: Open treatment of rib fracture(s) with internal fixation, including thoracoscopic visualization when performed, unilateral; 1–3 ribs [3].

21812: Open treatment of rib fracture(s) with internal fixation, including thoracoscopic visualization when performed, unilateral; 4–6 ribs [3].

21813: Open treatment of rib fracture(s) with internal fixation, including thoracoscopic visualization when performed, unilateral; 7 or more ribs [3].

> All three of these procedures are bundled with thoracoscopic visualization and are described by the Coders' Desk Reference as follows: Make a small incision over the displaced rib, dividing the skin, subcutaneous tissue, and muscle and exposing the ribs. Deflate the lung with the dual lumen endotracheal tube. Place a thoracoscopy through a small separate incision in the pleural cavity, and visualize the area of the fractured ribs. Make additional similar small skin incisions over the fracture sites as identified by thoracoscopic visualization.
>
> Identify and separate the intercostal nerve bundles. Using thoracoscopic visualization, fit the drill guide and plate or splint to the rib overlying the fracture. Drill the screw holes for attachment of the plate or splint insertion in each fracture site of the rib. Secure the splint or plate to the rib with locking screws to fix the hardware in place over the fracture. Repeat this procedure at each rib and/or fracture site. Remove the thoracoscopy, inflate the lung, and close the wounds in layers [4].

Additionally, each of these procedures is considered bundled, or falls into NCCI edits, restricting use with other procedure codes. Specifically, diagnostic video-assisted thoracoscopic surgery (VATS) (32601) is bundled with this procedure, but more extensive pulmonary evaluation and hematoma evacuation are not. See below for more details on "unbundled" procedures.

Hardware Insertion Documentation

Specifics related to the insertion of hardware provide a sufficient description of the procedure performed (please see section "General Principles"). The details of the procedure and exposure are of utmost importance and must include the location of the incision on the thoracic wall as well as its trajectory and size. An example of such a description could read "A vertical thoracotomy incision from cranial to caudal in the mid axillary line from the fifth intercostal space to the ninth intercostal space was performed." Additionally, the layers of the chest wall traversed and all muscular levels transected or split should be described in depth. Once the chest is exposed, a detailed description of the fractures as they are found prior to repair is also valuable.

Description of the procedure of fracture reduction, plate positioning, and temporary and eventually final hardware securement is also required. Association for the Study of Internal Fixation (AO) principles should be followed for best operative technique and should be described in detail in the operative note. This is the principle of anatomic reduction of fracture, bicortical alignment of the fracture, and periosteal plate apposition properly aligned and spanning the fracture site. Description of length of the hardware, including fixation technique (screw, clip, etc.), the distance beyond the fracture site, and the depth of the fixation device, should be dictated in accordance with the manufacturer's recommendations and FDA approval for the device insertion.

Additional Common Procedures

Chest Procedures: Open

32100: Thoracotomy; with exploration [3] (Table 15.1)

> The physician opens the chest cavity widely to directly visualize and assess the organs and structures in the chest. Using a scalpel, the surgeon makes a long incision around the side of the chest between two of the ribs. The incision is carried through all of the tissue layers into the chest cavity. Rib spreaders are inserted into the wound and the ribs are spread apart exposing the lung, heart, and other structures… The chest cavity is explored and the anatomy visualized using a gloved hand and large gauze sponges. The surgical instruments are removed. A chest tube(s) may be used to provide drainage for the chest cavity. When the procedure is complete, the operative wound is closed by sutures or staples [4].

32110: Thoracotomy with control of traumatic hemorrhage and/or repair of lung tear [3].

> The physician opens the chest cavity widely to directly visualize and assess the organs and structures in the chest. Using a scalpel, the surgeon makes a long incision around the side of the chest between two of the ribs. The incision is carried through all of the tissue layers into the chest cavity. Rib spreaders are inserted into the wound and the ribs are spread apart exposing the lung, heart, and other structures… The chest cavity is explored and the anatomy visualized using a gloved hand and large gauze sponges. The site of the hemorrhage or lung tear is identified and repaired. The surgical instruments are removed. A chest tube(s) may be used to provide drainage for the chest cavity. When the procedure is complete, … the operative wound is closed by sutures or staples [4].

32150: Thoracotomy with removal of intrapleural foreign body or fibrin deposit [3].

> The physician opens the chest cavity widely to remove a foreign body or fibrin deposit (thick insoluble protein deposit formed after the clotting of blood). Using a scalpel, the surgeon makes a long incision around the side of the chest between two of the ribs. The incision is carried through all of the tissue layers into the chest cavity. Rib spreaders are inserted into the wound and the ribs are spread apart exposing the lung. …Space is made in the chest by packing the uninvolved lung away from the operative field using large moist gauze sponges. The foreign body or fibrin deposit is located and removed by sharp and blunt dissection. When the procedure is complete, the instruments and gauze sponges are removed. A chest tube(s) may be used to provide drainage for the chest cavity. The operative wound is closed by sutures or staples [4].

32220: Thoracotomy with decortication, pulmonary (separate procedure); total [3].
32225: Thoracotomy with partial decortication [3].

> The physician removes a constricting membrane or layer of tissue from the surface of the lung (decortication) in order to permit the lung to fully expand. The physician opens the chest cavity widely. Using a scalpel, the surgeon makes a long incision around the side of the chest between two of the ribs. The incision is carried through all the tissue layers into the chest cavity. Rib spreaders are inserted into the wound and the ribs are spread apart exposing the lung. The constricting membrane is stripped off the surface of the lung. In 32225, only a portion of the lung surface is removed. The chest wall incision is sutured closed in layers. A chest tube(s) may be used to provide drainage for the chest cavity. When the procedure is complete, the skin is closed by suturing [4].

All of the above procedures that incorporate entering the thoracic cavity are not bundled in the NCCI edit for open reduction and internal fixation of the ribs; however they are all bundled with tube thoracostomy (32551), and this should not be reported.

Chest Procedures: Thoracoscopic

32601: Thoracoscopy, diagnostic (separate procedure), lungs, pericardial sac, mediastinal or pleural space, without biopsy [3].

> The physician examines the inside of the chest cavity through a rigid or flexible fiber optic endoscope. The procedure can be done under local or general anesthesia. The surgeon makes a small incision between two ribs and by blunt dissection and the use of a trocar enters the thoracic cavity. The endoscope is passed through the trocar and into the chest cavity. The lung is usually partially collapsed by instilling air into the chest through the trocar, or if general anesthesia is used, the lung may be collapsed through a double lumen endotracheal tube inserted through the mouth into the trachea. The contents of the chest cavity are examined by direct visualization and/or the use of a video camera. Still photographs may be taken as part of the procedure. At the conclusion of the procedure, the endoscope and the trocar are removed. A chest tube for drainage and re-expansion of the lung is usually inserted through the wound used for the thoracoscopy [4].

This is a bundled code and thus is considered a NCCI edit and should not be reported with any open reduction and internal fixation code (21811, 21812, 21813).

32651: Thoracoscopy with partial pulmonary decortication [3].

32652: Thoracoscopy with total pulmonary decortication including intrapleural pneumolysis [3].

The physician examines the inside of the chest cavity through a rigid or flexible fiber optic endoscope and removes a portion of the tissue covering the surface of the lung. The procedure can be done under local or general anesthesia. The physician makes a small incision between two ribs and by blunt dissection and the use of a trocar enters the thoracic cavity. The endoscope is passed through the trocar and into the chest cavity. The lung is usually partially collapsed by instilling air into the chest through the trocar or, if general anesthesia is used, the lung may be collapsed through a double lumen endotracheal tube inserted through the mouth into the trachea. The contents of the chest cavity are examined by direct visualization and/or by the use of a video camera. Still photographs may be taken as part of the procedure. A second and/or third trocar and instruments may be inserted into the chest cavity through a second and/or third wound in the chest. Under direct visualization through the endoscope, the physician strips away the membranous tissues covering a portion of the lung (or all of the lung ins 32652) using instruments inserted into the chest through the secondary sites. Code 32652 includes intrapleural pneumonolysis in which the physician divides the tissues attaching the lung to the wall of the chest cavity. At the conclusion of the procedure, the endoscope and the trocar are removed. A chest tube for drainage and re-expansion of the lung is usually inserted through the wound used for the thoracoscopy [4].

32653: Thoracoscopy with removal of intrapleural foreign body or fibrin deposit [3].

The physician examines the inside of the chest cavity through a rigid or flexible fiber optic endoscope and removes a foreign body or a fibrin deposit (the thick tissue much like remains of a blood clot). The procedure can be done under local or general anesthesia. The surgeon makes a small incision between two ribs and by blunt dissection and the use of a trocar enters the thoracic cavity. The endoscope is passed through the trocar and into the chest cavity. The lung is usually partially collapsed by instilling air into the chest through the trocar or, if general anesthesia is used, the lung may be collapsed through a double lumen endotracheal tube inserted through the mouth into the trachea. The contents of the chest cavity are examined by direct visualization and/or by the use of a video camera. Still photographs may be taken as part of the procedure. A second and/or third trocar and instruments may be inserted into the chest cavity through a second and/or third wound in the chest. The foreign body or fibrin deposit is located and removed using instruments through the scope or the secondary sites. At the conclusion of the procedure, the endoscope and the trocar are removed. A chest tube for drainage and re-expansion of the lung is usually inserted through the wound used for the thoracoscopy [4].

32654: Thoracoscopy with control of traumatic hemorrhage [3].

The physician examines the inside of the chest cavity through a rigid or flexible fiber optic endoscope and controls bleeding from a wound to the chest. The procedure can be done under local or general anesthesia. The physician makes a small incision between two ribs and by blunt dissection and the use of a trocar enters the thoracic cavity. The endoscope is passed through the trocar and into the chest cavity. The lung is usually partially collapsed by instilling air into the chest through the trocar or, if general anesthesia is used, the lung may be collapsed through a double lumen endotracheal tube inserted through the mouth into the trachea. The contents of the chest cavity are examined by direct visualization and/or by the use of a video camera. Still photographs may be taken as part of the procedure. A second and/or third trocar and instruments may be inserted into the chest cavity through a second and/or third wound in the chest. Under direct visualization through the endoscope, the physician manipulates the instruments inserted through the secondary sites and localizes the site of the bleeding. The hemorrhage is controlled by clipping or cauterizing the damaged blood vessel. At the conclusion of the procedure, the endoscope and the trocar(s) are removed. A chest tube for drainage and re-expansion of the lung is usually inserted through the wound used for the thoracoscopy [4].

As with the open thoracotomy codes, all thoracoscopic approaches tube thoracostomy (32551) are a bundled (NCCI edit) procedure and therefore not reported.

Additional Add-On Codes

32551: Chest tube insertion [3].

The physician removes fluid and/or air from the chest cavity by puncturing through the space between the ribs. To enter the chest cavity, the physician passes a trocar over the top of a rib, punctures through the chest tissues between the ribs, and enters the pleural cavity. Separately reportable imaging guidance may be used. With the end of the trocar in the chest cavity, the physician advances the plastic tube into the chest cavity. The sharp trocar is removed leaving one end of the plastic catheter in place within the chest cavity. A large syringe is attached to the outside end of the catheter and the fluid (blood or pus) is removed from the chest cavity by pulling back on the plunger of the syringe. The outside end of the tube may be connected to a drainage system, such as a water seal, to prevent air from being sucked into the chest cavity and to allow continuous or intermittent removal of air or fluid [4].

31622: Bronchoscopy, rigid or flexible, including fluoroscopic guidance, when performed; diagnostic, with cell washing, when performed (separate procedure) [3].

31624: Bronchoscopy with bronchial alveolar lavage [3].

The physician views the airway using a flexible fiberoptic or rigid bronchoscope that is introduced through the nasal or oral cavity. The airway is anesthetized. The bronchoscope is inserted and advanced through the nasal or oral cavity, past the larynx to inspect the bronchus….In 31624, the bronchoscopy includes bronchial alveolar lavage, which allows lung tissue to be sampled by irrigating with saline followed by suctioning the fluid. The bronchoscope is removed. These codes include fluoroscopic guidance, when performed [4].

31645: Bronchoscopy with therapeutic aspiration of tracheobronchial tree, initial (e.g., drainage of lung abscess) [3].

The physician views the airway using a bronchoscope introduced through the nasal or oral cavity, using local anesthesia of the patient's airway. The physician uses the views obtained through the bronchoscope to identify the closest approach to the fluid collection or abscess from within the airway. The physician passes a catheter through a channel in the broncho-

Table 15.1 CCI edits for 2181X codes

CPT code	21811	21812	21813	32551
32100	No edit	No edit	No edit	NCCI
32110	No edit	No edit	No edit	NCCI
32150	No edit	No edit	No edit	NCCI
32220	No edit	No edit	No edit	NCCI
32601	NCCI	NCCI	NCCI	No edit
32651	No edit	No edit	No edit	NCCI
32652	No edit	No edit	No edit	NCCI
32653	No edit	No edit	No edit	NCCI
32654	No edit	No edit	No edit	NCCI
32551	No edit	No edit	No edit	
31622	No edit	No edit	No edit	
31624	No edit	No edit	No edit	
31645	No edit	No edit	No edit	

From https://www.cms.gov/Medicare/Coding/NationalCorrectCodInitEd/index.html

> scope and aspirates the tracheobronchial tree. Alternately, the physician passes a needle through a channel in the bronchoscope into the fluid collection and aspirates fluid through the needle. The bronchoscope is removed [4].

The Elusive 22-Modifier

As noted earlier in the chapter, there are descriptive processes for determining the procedure performed. However, as every surgeon is aware, at times there are variables that can have a major impact on the difficulty of a case. The 22-modifier is therefore added to a CPT code to "up code" or increase the reimbursement as compensation for the increasing difficulty of the case. It is imperative that your documentation clearly describes where the difficulty was encountered, the extent of this difficulty, and how the procedure has varied from that described by the standard code. The AMA CPT code description of the 22-modifier is:

> ***Increased Procedural Services***: *When the work required to provide a service is substantially greater than typically required, it may be identified by adding modifier 22 to the usual procedure code. Documentation must support the substantial additional work and the reason for the additional work (i.e. increased intensity, time, technical difficulty of procedure, severity of patient's condition, physical and mental effort required)* [3].

It is therefore the responsibility of the operative surgeon to document in their detailed operative report not only the difficulty but also the increased amount of time that was required when compared to a standard operation relating to the base CPT code. Many centers have chosen to not submit for 22-modifier as this can lead to a higher rate of initial denial of claims. However a recent evaluation from the American College of Surgeons (ACS) [5] has demonstrated that even taking into account a claim denial and potential delay in reimbursement, there is financial benefit when the 22-modifier is used. Please see "Managing Denials" section for recommendations on denial of claims.

Managing Denials

Despite superb documentation and compliance with coding regulations, inevitably there will be denials of claims. Many private health-care organizations and reimbursement groups consider surgical stabilization of rib fractures, a new and emerging technology without sufficient data to support patient benefit. Frequently with denials of claims, denying agencies will allow for additional supporting evidence to be submitted along with the claim. Provided in the prior portions of this chapter, including documentation with description and appropriate NCCI edits that have been followed, it is likely that the denial of claim is a result of "insufficient evidence to support patient benefit." Many institutions have now created letters written by experienced clinicians that support the decision to perform such a procedure. Along with this letter, supporting literature with a bibliography as well as printed copies of these peer-reviewed journal articles is submitted along with the documentation and claims. In many cases, this additional work, while not an insignificant effort, has led to a reversal of denial and eventual payment.

References

1. Joint Commission Resources. https://www.jcrinc.com/store/publications/manuals/
2. https://www.cms.gov/Medicare/Coding/NationalCorrectCodInitEd/index.html?redirect=/NationalCorrectCodInitEd/
3. American Medical Association. Current procedural terminology. 4th ed. Chicago, IL: American Medical Association; 2016.
4. Coders' Desk Reference for Procedures: Answers to your toughest CPT coding questions.1st ed: OPTUMInsight;2016.
5. Bulletin, American College of Surgeons, Mar 2014, Gundersen Health System studies effect of modifier 22 on reimbursement for complex operations.

16 Future Directions

Fredric M. Pieracci

The present is an exciting time in the management of chest wall injuries. Prior to this decade, the approach to such injuries had been somewhat nihilistic, centered upon support of respiratory function using mechanical ventilation as well as analgesia achieved primarily with narcotics. These strategies have unfortunately lead to the frequent and iatrogenic complications of pneumonia and prolonged respiratory failure in the case of the former and substance abuse and addiction in the case of the latter [1].

The recognition of the dangers of both prolonged mechanical ventilation and narcotic use has fueled a burst of ingenuity surrounding the management of patients with chest wall injuries. Furthermore, advances in technology have allowed for the creation of rib-specific fixation systems, providing a tailored operative option for select patients with the most severe chest wall injuries. Finally, healthcare providers involved in the management of rib fracture patients have broadened to include additional surgical disciplines, such as orthopedic and thoracic surgery, as well as physical therapists, anesthesiologists, rehabilitation medicine physicians, respiratory therapists, and pulmonologists.

Many of these advancements, however, remain in their infancy and are still met with resistance from practitioners rooted in the aforementioned traditional mentality. Organization and cohesion of the rapidly developing knowledge surrounding rib fractures remains a challenge. This chapter will provide an agenda for the synthesis and maturation of the field of chest wall injury over the next decade, summarized in Table 16.1.

F. M. Pieracci
Department of Surgery, Denver Health Medical Center, University of Colorado School of Medicine, Denver, CO, USA
e-mail: Fredric.pieracci@dhha.org

M. de Moya, J. Mayberry (eds.), *Rib Fracture Management*,
https://doi.org/10.1007/978-3-319-91644-6_16

Table 16.1 Summary of future directions in the management of patients with rib fractures

Category	Task
Triage and prognostication	Refine and validate existing triage pathways for rib fracture patients, with incorporation of clinical variables
	Develop and validate a standardized nomenclature for rib fractures
	Refine scoring systems to incorporate complex fracture patterns
	Standardize or abandon the term "flail chest"
	Decrease or eliminate routine opioid prescription to patients with rib fractures
Operative indications	Incorporate and emphasize clinical variables into indications for surgery
	Evaluate the efficacy of SSRF in patients without flail chest
	Predict the progression of rib fracture displacement
	Evaluate the efficacy of clavicle and scapula repair in the setting of ipsilateral severe rib fractures
Operative technique	Develop pre-contoured posterior and anterior rib plates
	Develop rib-specific absorbable plates and screws
	Develop the tools for and technique of thoracoscopic SSRF
Systems	Develop and implement benchmarking for privileging in chest wall surgery
	Regionalize chest wall injury centers
	Create an international data repository
	Standardize and report complications of surgery
	Routinely measure and report long-term outcomes in studies of patients with rib fractures
	Develop and validate rib fracture-specific quality of life questionnaires

Identification of Rib Fracture Patients at Risk for Complications

Rib fractures encompass a wide range of injuries, from a single non-displaced fracture to multiple comminuted fractures series with associated bone loss. Furthermore, nearly identical fracture patterns may vary substantially in their phenotype based upon patient demographics, comorbidities, and coexisting injuries. Accordingly, there exists a substantial potential for mistriage of patients who present to the healthcare system. Persistent uncertainty regarding the disposition of rib fracture patients is in part due to an incomplete understanding of the variables that are most important in predicting adverse outcomes.

Only the minority of rib fracture patients develop life-threatening complications [2]; the challenge that remains is identifying them early. Many high-volume centers have developed a triage protocol for rib fracture patients (Fig. 16.1). However, there are few data available evaluating the predictive ability of such protocols and the relative contributions of individual variables [3, 4]. Currently available rib fracture scoring systems are based in large part upon a combination of radiographic and demographic variables (e.g., number of fractures, age) and do not incorporate physiologic variables [5]. Whereas general parameters such as the total number of ribs

RIB FRACTURE ALGORITHM

STEP 1

Admission if, after pain control with PO medications, ≥ 2 of these:

- ☐ RR > 18
- ☐ IS < 75% predicted*
- ☐ Numeric pain score ≥ 6
- ☐ Poor cough
- ☐ Age ≥ 65 years
- ☐ ≥ 3 fractures

STEP 2

if admission to hospital, SICU admission if ≥ 1 of these:

- ☐ Age ≥ 65 years
- ☐ ≥ 6 fractures
- ☐ IS < 60% predicted*
- ☐ Flail chest
- ☐ > 4 L O2 required to maintain sat > 90%
- ☐ Associated injury requiring ICU admission
- ☐ EM/surgery staff discretion

*As determined by EM respiratory therapist

PO (outpatient) pain regimen:

- Incentive spirometer
- Ibuprophen 800 mg PO q6H
- Gabapentin 100 mg PO TID
- Diazepam 5-10 mg PO QD
- Oxycodone PRN

Inpatient pain regimen:

- Incentive spirometer
- Ambulation
- Pain catheter vs. epidural
- Ibuprophen 800 mg PO q6H
- Gabapentin 100 mg PO TID
- Diazepam 5-10 mg PO QD
- Oxycodone PRN
- IV narcotics PRN
- Consider SSRF

Evaluate for SSRF if:

- Flail chest
- ≥ 3 severely displaced fractures
- ≥ 30% volume loss hemithorax
- Failure of non op management

Fig. 16.1 Denver Health Medical Center clinical pathway for patients who present to the emergency department with rib fractures. SSRF, surgical stabilization of rib fractures

fractured do predict crude outcome measures such as mortality [6], they fail to capture the majority of variability in outcomes.

Our group attempted to incorporate more detailed fracture pattern information into a rib fracture scoring system, entitled the RibScore, which included both the degree of fracture displacement and fractures in multiple locations (i.e., anterior, lateral, and posterior) [5]. Although the RibScore did possess favorable predictive ability, a substantial proportion of variability in respiratory outcomes was still unexplained. Therefore, it is necessary to evaluate more detailed fracture pattern information, such as consideration of subdivision by anatomic location (e.g., anterior vs. lateral vs. posterior) and relationship to other bones of the chest wall (e.g., clavicle, scapula).

This notion is contingent upon the development of standardized nomenclature to describe chest wall injuries, including fracture location, degree of displacement, and flail chest. For example, although the term "lateral" is used frequently to describe a rib fracture, it is unclear if this anatomic region is defined by surface anatomy (e.g., between the anterior and posterior axillary lines), radiographically (e.g., between one third and two thirds the distance from the sternum to the spine), or some combination of the two. Ambiguity in definitions limits any conclusions that may be drawn from studying subgroups of patients with certain rib fracture patterns (e.g., lateral rib fractures).

Another fundamental issue related to nomenclature involves the use (and frequent misuse) of the term “flail chest.” Although this term was used originally to describe the paradoxical motion of a floating segment of chest wall observed during physical exam [7], its use has expanded to refer to the radiographic finding of two or more adjacent ribs fractured in two or more locations. Beyond the ambiguity created by interchanging clinical and radiographic terms, defining flail chest radiographically creates a group of rib fracture patients with tremendous variability in clinical presentation and thus approach to management. There exists such a large physiologic variability in patients with “radiographic” flail chest that selection of this term as an inclusion criteria for either clinical pathways or research study risks the abstraction of nearly meaningless information. Unfortunately, it is precisely the use of radiographic flail chest that has driven inclusion criteria for most research involving patients with severe chest wall injuries, including scoring systems, analgesic regimens, and surgical stabilization of rib fractures (SSRF).

Although the term flail chest has been ingrained in the vernacular of the trauma surgeon, many chest wall surgeons, including the author, believe that it should be abandoned. Short of a moratorium, the term should be revised to clearly differentiate the clinical phenomenon (e.g., “flail segment”) from the radiographic finding (e.g., “floating segment”). Furthermore, caution should be exercised in using the ambiguous term “flail chest” as a research or clinical pathway inclusion criteria.

Several organizations are currently working collaboratively to develop a standardized nomenclature surrounding chest wall injuries, including the Chest Wall Injury Society (CWIS) and the AO Thoracic Expert Group. This system would ideally mirror other American Association for the Surgery of Trauma grading systems, which are characterized by objective, reproducible parameters [8]. Meaningful research into subgroup of patients with rib fractures will not be possible until such a system is developed and validated.

Another important future direction in the field of rib fractures will be the development and validation of high-fidelity scoring systems that extend beyond the relatively simplistic variables of fracture number and demographics. Parameters that require further evaluation include pulmonary physiologic variables such as respiratory rate, oxygen requirement, incentive spirometry, and cough ability. Dynamic parameters, such as progression of both symptoms and narcotic requirements over time, also warrant investigation. Some preliminary work in this arena has already laid the foundation for future large-scale trials [9–11].

Analgesia

Pain from rib fractures limits respiratory effort, resulting in pooling of secretions, atelectasis, and an increased risk of respiratory failure. Effective analgesia is particularly challenging for rib fractures, as compared to other bones, since the former cannot be immobilized effectively beyond operative fixation due to obligatory movement during respiration. Disabling pain from rib fractures is not limited to the acute period and may be permanent [12].

The vast majority of rib fracture patients are prescribed narcotics, irrespective of fracture number or pattern [1]. Fortunately, due to recent media attention, patients and providers have become more aware of the deleterious effects of unbridled opioid prescription [13]. One major shift in the management of rib fracture patients must be away from obligatory narcotics to control pain.

A myriad of nonnarcotic analgesics are available for the treatment of rib fracture pain. However, such modalities remain substantially underutilized. Patients presenting to non-trauma hospitals are particularly vulnerable to this phenomenon, as they are significantly less likely to receive locoregional anesthesia at the expense of narcotic prescription [14]. Novel analgesic approaches to rib fracture pain that require further investigation include oral gabapentin, intravenous ketamine and lidocaine infusions, and locoregional nerve blocks. This last category encompasses a wide range of techniques and target structures, including ultrasound-guided serratus and paravertebral blocks, percutaneous indwelling tunneled catheters, and thoracoscopic injection of liposomal bupivacaine into the subpleural space [15].

Improvement in pain control for patients with rib fractures, as well as a decrease in the use of narcotics to treat this pain, will likely require a multidisciplinary approach. Collaboration with pain management teams, including anesthesiologists and palliative care physicians, to develop rib fracture-specific analgesic pathways, will maximize implementation. Standardized analgesic regimens and assessment of outcomes will facilitate research into the relative merit of one approach over the other. Finally, outcomes assessed during investigations of rib fracture patients must routinely include both pain and narcotic requirements [11], as well as extend beyond the index admission.

Surgical Stabilization of Rib Fractures

Perhaps the most publicized example of an advancement in the field of chest wall injuries has been the development of SSRF. The proliferation of this operation is readily evident in the number of presentations at recent national meetings, the number of publications in peer-reviewed journals, and the number of new industry products that have come to market in the last 5 years [16]. Many believe that SSRF has been promulgated prematurely and in the absence of strong evidence. Cynics argue that it is the direct result of both influence from industry and the creation of SSRF-specific Common Procedural Terminology codes. Regardless of the reasons for the observed growth, as well as its ultimate efficacy, the age of SSRF is upon us, and the onus rests upon surgeons to maintain rigorous evaluation of this operation moving forward.

Indications for and Timing of Surgery

Henceforth, trials of SSRF have been limited mostly to patients with a radiographic diagnosis of flail chest and are characterized by a high degree of variability in patient selection, time to surgery, and method of follow-up [17–19]. One of the most pressing deficiencies in the SSRF literature involves the efficacy of the operation in patients with fracture patterns other than flail chest; this limitation has been recognized in national consensus statements [20, 21]. Despite these recommendations, our analysis of data from the National Trauma Database indicated that the majority of SSRF procedures are currently being done in patients without a diagnosis of flail chest [16].

Although there is theoretical benefit to stabilizing the chest wall in the presence of non-flail, severe chest wall injuries (e.g., multiple, bicortical, displaced fractures), this subgroup of patients has not been studied specifically. Rather, such patients have represented relatively small subgroups of patients within larger trials [22, 23]. We recently surveyed members of the CWIS as to their indications for SSRF in patients without flail chest and found a patient population in whom surgery was recommended approximately half of the time [24]. This hypothetical patient was used to generate inclusion criteria for a multicenter randomized trial which will investigate specifically the efficacy of SSRF in patients without flail chest.

Use of the relatively generic fracture patterns of both flail chest and fracture displacement likely underappreciates the complexity of chest wall injuries. The degree of disability that results from a flail segment likely depends upon additional variables such as the precise location of the fracture series on the ribs (e.g., posterior, lateral, or anterior), the number of ribs involved, and the position of the segment in a superior-inferior plane. Furthermore, fracture displacement is a dynamic variable; certain patients who present with non-displaced fractures may subsequently progress to severe, bicortical displacement (Fig. 16.2). Identifying patients at risk for this progression remains a challenge. Additional fracture patterns that require further investigation include rib fractures in conjunction with either clavicle or scapular fractures, subscapular fractures, and anterior flail segments as a result of chest compressions. Conversely, the conditional efficacy of both clavicle and scapula repair in the setting of an ipsilateral chest wall injury requires specific investigation.

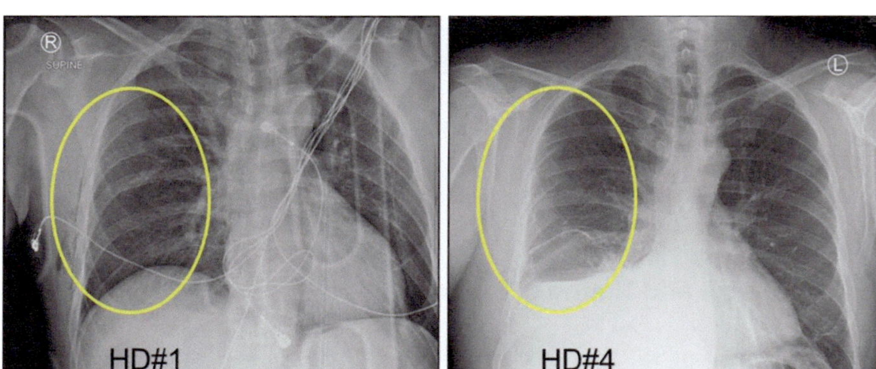

Fig. 16.2 Interval displacement of posterior rib fractures. *HD* hospital day

Another limitation of currently proposed indications for surgery is the failure to incorporate pulmonary physiologic variables. Many surgeons will have anecdotes related to patients with severe radiographic chest wall injuries who are clinically well; conversely, patients with one or two displaced rib fractures may be bothered with a painful clicking sensation that resolves with surgical fixation. Examples of pulmonary parameters include incentive spirometry, vital capacity, numeric pain score, cough quality, and analgesic requirements. Efforts have been made to utilize bedside limited pulmonary function testing to identify patients at high risk for failure of nonoperative management [9, 10, 25]. Standardization of the aforementioned pulmonary physiologic variables into study inclusion criteria represents an important charge for the future.

In addition to indications for surgery, the optimal timing of SSRF remains unknown. Beyond clear contraindications to surgery, such as shock, intracranial hypertension, and unstable pelvis fractures, it is unclear if earlier surgery confers a benefit as compared to later surgery. Proponents of early surgery argue that certain clinical and radiographic variables, present on or close to admission, can predict with a high degree of certainty failure of nonoperative management. By contrast, advocates of delayed surgery argue that a trial of nonoperative management will allow certain patients to "declare" themselves as candidates for surgery.

In a multicenter analysis of prospectively collected SSRF databases, we found that surgery within 24 h of admission was associated with favorable pulmonary outcomes as compared to surgery between 3 and 10 days after admission. These data represented highly selected patients from high-volume centers, and additional prospective data are necessary to refine the optimal timing of SSRF.

Barriers to Surgery

In a now 10-year-old survey, surgeons indicated lack of both expertise and prospective research as the most important barriers to offering SSRF to their patients [26]. Since that time, multiple additional trials related to SSRF have been published, and most national organizations, as well as third-party vendors, offer courses on surgical technique. Short of the aforementioned survey, there exists a very rudimentary understanding of the number of patients with severe chest wall injuries who do not undergo surgery and why. To be sure, this number is a moving target secondary to a lack of firm indications for surgery. However, there exists a subset of patients who are offered surgery and decline. Understanding the reasons for their declination will aid in refining the benefit of the operation, as well as developing education materials for both patients and providers. Moreover, a second subset of patients is deemed ineligible for surgery due to both pulmonary and non-pulmonary coexisting injuries. Both groups require further investigation. Finally, the likelihood of being offered SSRF for flail chest is significantly associated with presentation to centers performing this operation with regularity [16]. To the extent that it is possible, efforts should be made to minimize discrimination of patients with identical injuries either for or against SSRF.

Operative Technique

The earliest attempts at operative repair of rib fractures involved either external or internal fixation with materials intended for bones with properties vastly different from those of the ribs [27]. Examples included K wires and mandibular plates. A major development in the technique of SSRF occurred in the 1990s with the development of rib-specific plates and struts [28]. The development of these systems was fostered by an appreciation of the unique biomechanical properties of the ribs as compared to other long bones, such as contouring and elasticity.

Despite these advances, several challenges remain. Pre-contoured plates are matched to the lateral portion of the ribs, such that apposition to fractures at the rib margins requires a substantial degree of intraoperative bending. This problem is particularly evident in posterior rib fractures, where the acute angle of the rib, in conjunction with pulling forces from the serratus anterior muscle, results in a high rate of hardware failure [29]. The availability of a high-fidelity, pre-contoured posterior rib plate remains a gap in the surgeon's armamentarium.

Anterior fractures present a different set of challenges due to their proximity to both costal cartilage and the sternum. Additional research is necessary to evaluate the long term integrity of fixation to cartilage. One important development in the surgical management of anterior fractures has been the creation of the rib- and sternum-specific "T plates," which provide rigid fixation to both these structures (Fig. 16.3).

One final development consideration regarding surgical technology is that of absorbable plates. Although uncommon, infection of permanent hardware used in SSRF is highly morbid [30]. Implantation of permanent hardware remains a significant barrier to both surgeons and patients in terms of acceptance of surgery [26].

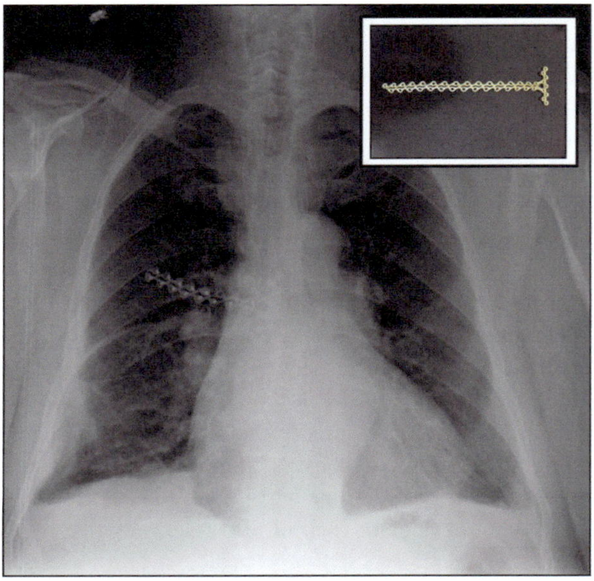

Fig. 16.3 "T plate" for fixation of the sternum to anterior rib fracture fragment

There is currently no rib-specific absorbable plating system available in the USA. However, favorable data in both animals [31] and humans [32] have provided the basis for the development of such as system. The study of absorbable plates and screws for either routine or selective using during SSRF represents a necessary progression in the field of chest wall surgery.

Irrespective of the fixation system employed, substantial progress has also been made in both positioning and exposure of rib fractures for repair. Until recently, almost all SSRF operations were accomplished in the lateral decubitus position, via a traditional posterolateral thoracotomy incision which paralleled the course of the underlying ribs. The latissimus dorsi, trapezius, rhomboids, and serratus muscles were routinely divided, and scapular retraction was the norm. Such maneuvers resulted in a considerable amount of postoperative and long-term morbidity.

As elective thoracic surgery has undergone a transition toward minimally invasive techniques [33], so too has SSRF. Experience with the procedure has grown, and many surgeons now employ relatively small incisions and muscle retraction in lieu of division. One specific development in the author's technique has been a shift from a horizontal to a vertical incision, centered along the medial border of the latissimus dorsi muscle, for approaching lateral fracture series (Fig. 16.4) [34]. Such an incision allows retraction rather than division of the latissimus muscle, as well as exposure to the lateral ribs 3–8 via a single split of serratus fibers.

Creative approaches to both anterior and posterior fractures have also been described. For example, many surgeons now employ the prone position to access posterior fracture series, which serves to lateralize the scapula (rather than retract it), and minimize muscle division via development of the triangle of auscultation. Prone positioning also allows for concomitant repair of thoracic spine fractures (Fig. 16.5). The development of powered, right-angle drills and screwdrivers has further facilitated exposure to fractures through relatively small incisions.

Perhaps the "final frontier" in minimally invasive SSRF involves completely thoracoscopic surgery, with placement of hardware on the inner cortex of the ribs. Such an approach has multiple advantages over traditional surgery, such as smaller incisions, access to the ribs at their margins, diagnosis and treatment of pleural pathology, and elimination of palpable subcutaneous plates. Although this operation has been performed [35], there remains a large gap between theory and technology. The operation remains limited by currently available plates (designed to be placed on the outer cortex of the rib), reduction tools, and drills/screwdrivers that lack the 180 articulation necessary to address the concavity of the thoracic cage thoracoscopically. This operation may be particularly amenable to the use of the robot.

Three-dimensional printing systems have provided a complementary technology to aid in operative planning for select chest wall deformities. Although this technology remains immature, there are multiple theoretical uses, including printing contralateral, normal ribs for pre-contouring plates in the case of severe fracture comminution. Such an approach could decrease both operative time and maximize plate fitting. Printing a 3D reconstruction of a patient's chest wall may also aid in informing and educating the patient as to the severity of their injury and the technicalities of the operation.

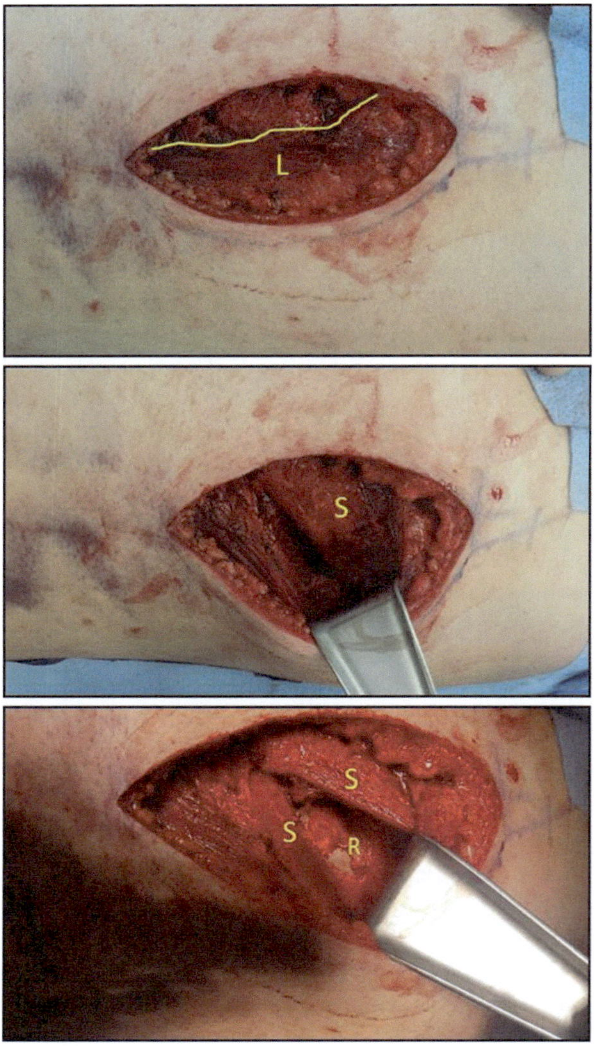

Fig. 16.4 Muscle sparring approach to lateral rib fracture series. *L* latissimus dorsi muscle, *S* serratus anterior muscle, *R* rib

Reporting and Treatment of Complications

Fortunately, serious complications of SSRF appear to occur relatively and rarely. However, because there is no standardized reporting system for such complication, the current literature consists of single-center reports of variably defined adverse events. Serious complications, such as hardware infection [30] and nerve injury [36], are only now beginning to be published. Guidance in this area is available from several other disciplines, for example, a graded scale for reporting bleeding complications in clinical evaluations of acute coronary syndrome [37]. Application of such a scale to the complication of hardware infection following SSRF might appear as

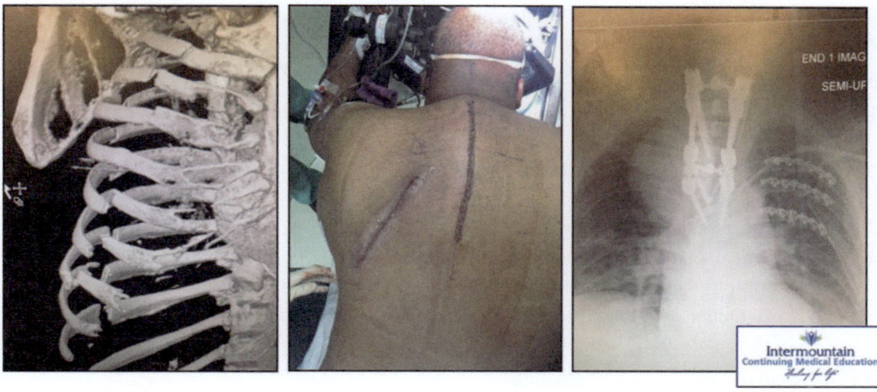

Fig. 16.5 Concomitant repair of posterior rib fracture series and thoracic spine fracture via the prone position

Table 16.2 Hypothetical example of a standardized definition and grading scale for reporting hardware infection following surgical stabilization of rib fractures

Definition	1. Evidence of local infection including one or more of the following: erythema, edema, tenderness at the surgical site, expression of pus from the surgical site, radiographic (CT or ultrasound) findings of enhancing or air-containing fluid surrounding the hardware 2. Clinical signs of systemic infection including two or more of the following: fever >38.5 °C, leukocytosis >12,000 k/uL, respiratory rate >20 breaths per minute, heart rate >110 beats per minute 3. In the case of sampled fluid, positive growth of one or more pathogens
Subgrouping	1. Early: within 14 days of surgery 2. Late: greater than 14 days from surgery
Grading	*Grade I*: skin changes that respond to antimicrobial therapy alone *Grade II*: infection requires opening of wound with or without drainage of infected fluid. Hardware remains in situ *Grade III*: infection requires removal of hardware

shown in Table 16.2. Standardized, universal definitions for surgical complications will enable meaningful comparisons to be made and ultimately improve patient outcomes.

Systems Issues

The last two decades have seen a surge in both the amount and detail of information surrounding chest wall injuries. Most surgeons involved in the management of these injuries agree that the amount of information has reached a level sufficient to warrant consideration of "chest wall injuries" as a unique field of expertise within trauma surgery. Accordingly, several systems issues that apply to other specialties are necessary to consider as this new field matures.

Table 16.3 Suggested minimum components for privileging in the surgical management of chest wall injuries

Number	Description
1	Board eligible or board certified in general surgery, orthopedic surgery, or thoracic surgery
2	Participating in a workshop/course teaching the technique of SSRF, with appropriate documentation
3	Preceptorship by a surgeon privileged in SSRF for 5–10 cases
4	Critical analysis of outcomes through participation in institutional morbidity and mortality conferences and/or submission of outcomes to a national data repository
5	Maintenance of expertise by performance of at least ten SSRF operations each 2-year period

SSRF surgical stabilization of rib fractures

Credentialing and privileging in the operative management of chest wall injuries represent an important step in this process. Although each institution's medical staff will differ in terms of the necessary requirements, centers interested in developing a chest wall injury program should develop privileging standards for both new and existing surgeons who will perform these operations. Ideally, a minimum set of standards would be promulgated by a national or international society, as has been done within other fields of surgery [38]. Standard minimum privileging elements to consider are listed in Table 16.3.

Ample data now exist documenting the relationship between volume and outcome over a wide range of surgical procedures [39]. However, there is currently no strong evidence that this relationship applies to SSRF. Our analysis of the NTDB for SSRF suggested that outcomes following SSRF were improved in Level I and higher-volume trauma centers; however, these data were likely confounded by additional institutional and regional parameters [16]. Continued refinements in the technique of SSRF, including thoracoscopic approaches, bone grafting, and management of fractures of the margin, will likely strengthen the argument for the creation of specialized referral centers for chest wall injured patients. Moreover, surgical expertise represents only a portion of the resources necessary for a robust program; additional considerations include pain management, physical therapy, and pulmonary rehabilitation teams. Evaluation of the outcomes of patients who present to such centers, as compared to other trauma and non-trauma centers, is necessary. Finally, SSRF should be considered as a new index case for incorporation into both general surgery residencies and trauma and acute care surgery fellowship.

Along the lines of the creation of specialty centers with expertise in the management of chest wall-injured patients, formation of both national and international societies, comprised of healthcare providers with interest and expertise in chest wall injuries, is needed to further advance the field. Such groups may represent subgroups within existing organizations (e.g., the American Association for the Surgery of Trauma) or de novo societies (e.g., the CWIS). These groups will be charged with setting the standards for privileging and credentialing, nomenclature and outcomes, clinical practice guidelines, maintenance of a data repository, education, networking, and conduct of research. Creation of an international data repository will provide a standardized means by which to compare outcomes across centers, management modalities, and outcome definitions. Such a database would ideally be housed within such an organization. Finally, such organizations should strive to include representatives from trauma, thoracic and orthopedic surgery, facilitating cross pollination of information and encouraging collaboration.

Outcomes

Population-based cross-sectional analyses have documented the high likelihood of prolonged and even permanent disability associated with severe chest wall injures [12, 40]. However, most therapeutic studies of interventions targeted to such patients have limited outcomes to the acute period. Moreover, relatively crude outcomes such as respiratory failure, tracheostomy, and mortality have been reported. As the proportion of trauma patients who die from chest wall injuries continues to decrease [41], a shift toward analysis of quality of life among survivors will be necessary.

An understanding of the impact of newer analgesic and surgical technologies on long-term quality of life is only beginning to be explored. Imperative to this process will be the development and validation of rib fracture-specific quality-of-life measurements. Consider, for example, long-term shoulder girdle disability secondary to grinding against a displaced posterior fracture series. Such an outcome has not been addressed prospectively in any study of rib fracture patients. By contrast, most investigations have employed relatively generic quality of life measurements [42, 43]. More recently, reports of SSRF-specific questions have been published [23]; however, these questionnaires require validation. Involvement of rib fracture patients themselves in the design and implementation of future trials will likely enhance further the quality of the research.

Conclusions

The field of chest wall injuries is developing rapidly. Recognition of the management of these injuries as a distinct skill set that encompasses sufficient knowledge per se is an important first step. Much of the foundation for the management of rib fractures has already been laid, albeit in a relatively disorganized fashion. In general, refinement of nomenclature, analgesia, and surgical indications represent the three most pressing areas requiring attention in the next decade. Fortunately, there is ample material with which to conduct research, and an eager investigator could find several research topics within this chapter alone. As the technology and publicity surrounding rib fractures, and in particular their surgical management, continues to expand, the challenge for clinicians will be to remain grounded in the basic principles of collaboration, objectivity, and hypothesis testing. There remains much room for improvement surrounding this common and highly morbid injury.

References

1. Alghnam S, Castillo R. Traumatic injuries and persistent opioid use in the USA: findings from a nationally representative survey. Inj Prev. 2017;23(2):87–92.
2. Majercik S, Pieracci FM. Chest wall trauma. Thorac Surg Clin. 2017;27(2):113–21.
3. Gonzalez KW, et al. A pilot single-institution predictive model to guide rib fracture management in elderly patients. J Trauma Acute Care Surg. 2015;78(5):970–5.
4. Pressley CM, et al. Predicting outcome of patients with chest wall injury. Am J Surg. 2012;204(6):910–3. discussion 913–4
5. Chapman BC, et al. RibScore: a novel radiographic score based on fracture pattern that predicts pneumonia, respiratory failure, and tracheostomy. J Trauma Acute Care Surg. 2016;80(1):95–101.
6. Flagel BT, et al. Half-a-dozen ribs: the breakpoint for mortality. Surgery. 2005;138(4):717–23. discussion 723–5
7. Landercasper J, Cogbill TH, Lindesmith LA. Long-term disability after flail chest injury. J Trauma. 1984;24(5):410–4.
8. American Association for the Surgery of Trauma Organ Injury Scoring Scales. http://www.aast.org/Library/TraumaTools/InjuryScoringScales.aspx. Assessed 12 Oct 2017.
9. Carver TW, et al. Vital capacity helps predict pulmonary complications after rib fractures. J Trauma Acute Care Surg. 2015;79(3):413–6.
10. Butts CA, et al. Do simple beside lung function tests predict morbidity after rib fractures? Am J Surg. 2017;213(3):473–7.
11. de Moya M, et al. Pain as an indication for rib fixation: a bi-institutional pilot study. J Trauma. 2011;71(6):1750–4.
12. Fabricant L, et al. Prolonged pain and disability are common after rib fractures. Am J Surg. 2013;205(5):511–5. discusssion 515–6
13. Glowacki EM, Glowacki JB, Wilcox GB. A text-mining analysis of the public's reactions to the opioid crisis. Subst Abus. 2017:1–5.
14. Gage A, et al. The effect of epidural placement in patients after blunt thoracic trauma. J Trauma Acute Care Surg. 2014;76(1):39–45. discussion 45–6
15. Truitt MS, et al. Continuous intercostal nerve blockade for rib fractures: ready for primetime? J Trauma. 2011;71(6):1548–52. discussion 1552
16. Kane E, et al. Quantifying and exploring the recent national increase in surgical stabilization of rib fractures. J Trauma Acute Care Surg. 2017;83(6):1047–52.
17. Leinicke JA, et al. Operative management of rib fractures in the setting of flail chest: a systematic review and meta-analysis. Ann Surg. 2013;258(6):914–21.
18. Coughlin TA, et al. Management of rib fractures in traumatic flail chest: a meta-analysis of randomised controlled trials. Bone Joint J. 2016;98-B(8):1119–25.
19. Slobogean GP, et al. Surgical fixation vs nonoperative management of flail chest: a meta-analysis. J Am Coll Surg. 2013;216(2):302–11. e1
20. Pieracci FM, et al. Consensus statement: surgical stabilization of rib fractures rib fracture colloquium clinical practice guidelines. Injury. 2017;48(2):307–21.
21. Kasotakis G, et al. Operative fixation of rib fractures after blunt trauma: a practice management guideline from the Eastern Association for the Surgery of Trauma. J Trauma Acute Care Surg. 2017;82(3):618–26.
22. Pieracci FM, et al. A prospective, controlled clinical evaluation of surgical stabilization of severe rib fractures. J Trauma Acute Care Surg. 2016;80(2):187–94.
23. Majercik S, et al. Long-term patient outcomes after surgical stabilization of rib fractures. Am J Surg. 2014;208(1):88–92.
24. Pieracci FM, et al. Indications for surgical stabilization of rib fractures in patients without flail chest: surveyed opinions of members of the Chest Wall Injury Society. Int Orthop. 2017;42(2):401–8.

25. Doben AR, et al. Surgical rib fixation for flail chest deformity improves liberation from mechanical ventilation. J Crit Care. 2014;29(1):139–43.
26. Mayberry JC, et al. Surveyed opinion of American trauma, orthopedic, and thoracic surgeons on rib and sternal fracture repair. J Trauma. 2009;66(3):875–9.
27. Bemelman M, et al. Historic overview of treatment techniques for rib fractures and flail chest. Eur J Trauma Emerg Surg. 2010;36(5):407–15.
28. Bottlang M, et al. Biomechanical rationale and evaluation of an implant system for rib fracture fixation. Eur J Trauma Emerg Surg. 2010;36(5):417–26.
29. Marasco SF, Sutalo ID, Bui AV. Mode of failure of rib fixation with absorbable plates: a clinical and numerical modeling study. J Trauma. 2010;68(5):1225–33.
30. Thiels CA, et al. Infected hardware after surgical stabilization of rib fractures: outcomes and management experience. J Trauma Acute Care Surg. 2016;80(5):819–23.
31. Vu KC, et al. Reduction of rib fractures with a bioresorbable plating system: preliminary observations. J Trauma. 2008;64(5):1264–9.
32. Mayberry JC, et al. Absorbable plates for rib fracture repair: preliminary experience. J Trauma. 2003;55(5):835–9.
33. Hartwig MG, D'Amico TA. Thoracoscopic lobectomy: the gold standard for early-stage lung cancer? Ann Thorac Surg. 2010;89(6):S2098–101.
34. Pieracci FM, et al. Surgical stabilization of severe rib fractures. J Trauma Acute Care Surg. 2015;78(4):883–7.
35. Pieracci FM, et al. Completely thoracoscopic, intra-pleural reduction and fixation of severe rib fractures. Trauma Case Reports. 2015;1:39–43.
36. Skedros JG, et al. Medial scapular winging associated with rib fractures and plating corrected with pectoralis major transfer. Int J Surg Case Rep. 2014;5(10):750–3.
37. Rao SV, et al. Standardized reporting of bleeding complications for clinical investigations in acute coronary syndromes: a proposal from the academic bleeding consensus (ABC) multidisciplinary working group. Am Heart J. 2009;158(6):881–886e1.
38. Inabnet WB III, et al. Joint task force recommendations for credentialing of bariatric surgeons. Surg Obes Relat Dis. 2013;9(5):595–7.
39. Birkmeyer JD, et al. Surgeon volume and operative mortality in the United States. N Engl J Med. 2003;349(22):2117–27.
40. Mayberry JC, et al. Long-term morbidity, pain, and disability after repair of severe chest wall injuries. Am Surg. 2009;75(5):389–94.
41. Dehghan N, et al. Flail chest injuries: a review of outcomes and treatment practices from the National Trauma Data Bank. J Trauma Acute Care Surg. 2014;76(2):462–8.
42. Marasco SF, et al. Prospective randomized controlled trial of operative rib fixation in traumatic flail chest. J Am Coll Surg. 2013;216(5):924–32.
43. Marasco S, et al. Quality of life after major trauma with multiple rib fractures. Injury. 2015 Jan;46(1):61–5.

Index

M. de Moya, J. Mayberry (eds.), *Rib Fracture Management*,
https://doi.org/10.1007/978-3-319-91644-6

MIX
Papier aus verantwortungsvollen Quellen
Paper from responsible sources
FSC® C105338

If you have any concerns about our products,
you can contact us on
ProductSafety@springernature.com

In case Publisher is established outside the EU,
the EU authorized representative is:
Springer Nature Customer Service Center GmbH
Europaplatz 3, 69115 Heidelberg, Germany

Printed by Libri Plureos GmbH
in Hamburg, Germany